New computational techniques for strongly correlated electron systems

Dissertation zur Erlangung des
naturwissenschaftlichen Doktorgrades
der Bayerischen Julius-Maximilians-Universität
Würzburg

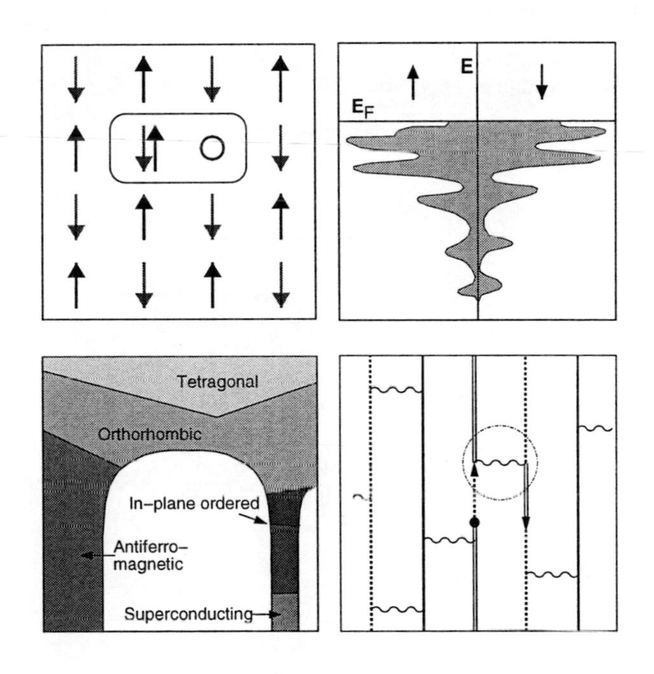

vorgelegt von

Ansgar Dorneich

aus

Freiburg

Würzburg 2001

Eingereicht am 12. Dezember 2001
bei der Fakultät für Physik und Astronomie

1. Gutachter: Prof. Dr. Werner Hanke
2. Gutachter: Priv. Doz. Dr. Enrico Arrigoni
der Dissertation

1. Prüfer: Prof. Dr. Werner Hanke
2. Prüfer: Prof. Dr. Rene Matzdorf
der mündlichen Prüfung

Tag der mündlichen Prüfung: 11. Februar 2002

Bibliografische Information Der Deutschen Bibliothek

Die Deutsche Bibliothek verzeichnet diese Publikation in der Deutschen Nationalbibliografie; detaillierte bibliografische Daten sind im Internet über http://dnb.ddb.de abrufbar.

ISBN 3-8325-0228-9

Logos Verlag Berlin
Comeniushof, Gubener Str. 47,
10243 Berlin
Tel.: +49 030 42 85 10 90
Fax: +49 030 42 85 10 92
INTERNET: http://www.logos-verlag.de

To err is human, but to really foul things up requires a computer.
Farmers' Almanac, 1978

Contents

Introduction

Strongly correlated electron systems rank amongst the most intensively studied objects in modern theoretical solid state physics [1, 2]. The reason for this interest is that strongly interacting electrons are the key ingredient for many exotic optical, electrical and magnetic properties of technologically promising materials such as high temperature superconductors (HTSC) [3] or colossal magnetoresistance materials (CMR) [4]. Amongst many others, possible applications for these classes of materials range from magnetic data recording to dramatically faster computers, loss-free transport and storage of electrical energy, new means of public transport based on magnetic levitation, medical imaging, better speech quality in mobile communication, and hyperfine detectors in scientific research and material testing.

The fundamental property of strongly correlated electron systems is an extraordinarily large entanglement of the many-body wave function of a "macroscopic" number, about 10^{23}, of electrons within them. This essentially "quantum mechanical behavior on a macroscopic level" causes unusual material properties such as resistance-free charge transport and ideal diamagnetism in the case of the HTSC. On the other hand, the same behavior makes a theoretical description extremely difficult. The standard analytical approach for the description of interacting many-particle systems is to start from the corresponding noninteracting systems and to add the interaction effects a posteriori as (hopefully) small perturbations or effective fields and potentials. Concepts of this kind have been applied very successfully almost throughout the modern theory of simple metals, insulators, or semiconductors [5, 6]. More generally, traditional theories in solid state physics rely on the existence of a small parameter such as the ratio between Coulomb interactions of the charge carriers and the kinetic energy, or bandwidth, in a typical metal. However, this approach, which implicitly assumes that the properties of the interacting system evolve 'smoothly' from those of the noninteracting one, must fail if the effect to be studied relies on the strong interaction of many particles. In the case of the HTSC, for example, this fundamental problem is expressed by the fact that the Coulomb interaction is equal to or even larger than the bandwidth.

For this reason, numerical simulations have become important tools within solid state physics, and a substantial part of the literature and ongoing works in this area rely on these 'computer experiments' [1, 2, 7]. Numerical simulation allows measurement of the

properties of well defined microscopic model Hamiltonians up to essentially arbitrary accuracy, thereby taking into account all quantum mechanical interactions without any approximation or perturbation expansion. Most practically important simulation techniques are able to calculate dynamic response and correlation functions, which permits "to simulate in principle any conceivable real experiment, but also to perform experiments which are not (yet) possible in real life" [2].
Numerical simulations are not the only field in which computational methods can be successfully applied to the problems of solid state physics, other promising applications are computer aided algebraic manipulations (CAAM). Computer algebra frameworks like Maple^{TM} or Mathematica^{TM} [8] allow the use of computers as highly efficient term manipulating machines for almost any grammatically closed and logically consistent formal rule system. A good example for the usefulness of such a computer algebra system in physics is the program package Feynarts [9] which calculates arbitrary Feynman diagrams up to a desired expansion order. For quantum mechanical many-particle systems in solid state physics, the formalism of Second Quantization [10] provides a logically consistent rule system with a relatively limited alphabet and grammar. A CAAM machine operating on this formal language can serve to find and characterize energy eigenstates, to perform basis transformations, or to construct quasiparticle descriptions of the low energy physics of a given Hamiltonian – a list which is in no way complete.
In this work, new approaches from both previously described disciplines of computational physics will be presented. Each new algorithm and code package will be followed by one or more applications to currently discussed problems in solid state physics. The following chapters are organized as follows:
The first chapter gives an overview on currently available computational methods for quantum mechanical many-particle systems. Some widely studied Hamiltonians and microscopical models are presented, the most important numerical simulation techniques are summarized and problems arising from the small accessible cluster size (finite-size effects) are discussed.
The following chapter is dedicated to Exact Diagonalization (ED), the most elementary numerical simulation technique in solid state physics. Key features, strengths and limitations of the method are described, and tricks for an efficient implementation are presented. The most important trick turns out to be a clever exploitation of the studied system's inherent symmetries, which correspond to certain invariances of the Hamiltonian. Other helpful concepts are matrix-free methods, approaches to reduce the enormous memory requirement of the ED technique which grows exponentially with the number of orbitals in the studied system. ED methods have been successfully applied to two main classes of physical systems. They have been applied on the one hand, to highly-ordered periodic arrangements of single orbitals, modeling small cuttings of perfectly grown crystal lattices. These small cuttings of not more than a few dozens of orbitals are often closed onto themselves by periodic boundary conditions (PBC) in order to artificially restore some features of the quasi-infinite lattice. On the other hand, ED methods have been used to simulate single unit cells of metal-ligand composites with relatively complex internal structures and many orbitals per unit cell; these unit cells are typically separated from their environment in the crystal lattice by applying open boundary conditions (OBC). These two classes of applications require different coding concepts and for each of them an exemplary implementation is described and presented.

At the end of the chapter, these two program packages are used to gain new insights into two currently discussed problems in theoretical solid state physics. By means of the code for periodic lattices we study the zero temperature phase diagram of the Hubbard model [11] (see chapter 1) and identify two different metal-insulator transitions. The first of these has an insulating state of Mott-Hubbard type, in which the insulating behavior is uniquely due to electrostatic electron-electron repulsion), the second one with an insulating state of Mott-Heisenberg type (in which an antiferromagnetic ordering of electron spins creates the insulating behavior). Parts of this work have been published in [12]. In a second application, the code for single unit cells is used to explain experimentally recorded X-ray magnetic circular dichroism spectra (XMCD) [13] of the technologically promising magnetic transition metal oxide CrO_2. It turns out that the experimental spectra can be reproduced with high accuracy by studying one elementary CrO_6 octahedron within the material (with oxygen atoms at the six corners and one chromium atom at the center of the octahedron). The ED studies demonstrate that ligand fields connected with cluster deformation (away from the ideal octahedron form) and spin-orbit coupling of angular momenta both in the valence shell and in the core shell of the chromium atom play a decisive role for reproducing the experimental spectra. This work will be published in [14].

Chapter 3 describes a recently proposed Quantum Monte Carlo technique, the Stochastic Series Expansion method (SSE) [16, 17]. Unlike many 'traditional' QMC techniques (see [18] for an overview), SSE is an exact method without any systematic error such as the Trotter discretization error. Furthermore, a non-local loop-update mechanism can easily be implemented within SSE [16]. This approach, which is similar to the loop algorithm [19], allows for much lower autocorrelations times on large systems or at low temperatures than any purely local update could reach [20]. Such a favorable scaling behavior makes it possible to simulate systems with several thousand lattice sites and/or systems at very low temperatures. Within this thesis, an object-oriented C++ implementation of the SSE technique has been created which considerably enhances the original concept by incorporating a very efficient way to measure arbitrary Green's functions within SSE [20]. Chapter 3 gives a short review on the basic concepts of SSE and then describes in detail the new methods for measuring Green functions. Additionally, benchmark calculations and a detailed analysis of SSE's scaling behavior as a function of system size and inverse temperature are presented.

In the object-oriented software design of the new SSE codes, much attention has been spent on the creation of a flexible program which can easily be applied to wide classes of lattice geometries and Hamiltonians. This has been achieved by strictly separating the numerical simulation routines from the program modules describing the crystal lattice and from the parts modeling the particles and their energetic interactions. A specific application then requires nothing but selecting a certain lattice geometry and a certain Hamiltonian and combining them with the general SSE simulation routines. The flexibility of this design is demonstrated in the second part of chapter 3, in which several exemplary studies based on SSE are presented. Some of these works have been published in more detail in separate publications [21, 22]. Another application, a detailed numerical study of the projected SO(5) model [23], is treated separately in chapter 5 of this work.

After two chapters on new developments in numerical simulation techniques, chapter 4 raises the question of how computer algebra can be useful in solving current problems in theoretical solid state physics. Starting from a few general considerations on design guidelines for a term manipulation machine on top of the formal language of Second Quantization, a Mathematica^{TM} package is presented which implements most of the functionality discussed above. As an exemplary application, the so-called Hubbard-I approximation [24] of the Hubbard model near half-filling is revisited and enhanced. In Hubbard's expansion, the Coulomb interaction between the electrons is treated exactly and the kinetic energy is added in an approximative form as a small expansion parameter. This approximation, which is valid in the limit $U/t \to \infty$, yields an energy spectrum of two distinct bands separated by the energy gap U. Recent QMC studies for the paramagnetic phase of the Hubbard model [25], on the contrary, show four bands in the spectral function instead of two. Also Exact Diagonalization studies [1] produce results which are not in accordance with the Hubbard-I approximation. This was the motivation to study the leading corrections to the Hubbard-I approximation. It turns out that the four-band structure in the QMC data can be reproduced quite well by introducing two composite quasiparticles. However, after introducing these quasiparticles, many extremely cumbersome commutator evaluations are neccessary to formulate the new equations of motion. These commutator evaluations can only be performed by an automated term manipulation system such as the one described above.

The last chapter of this work presents a detailed numerical study of the projected SO(5)-symmetric model of high-temperature superconductivity – the main physical research area of the author during this thesis. The projected SO(5) model (or pSO(5) model) [23] tries to combine the ideas of SO(5) symmetry between the antiferromagnetic and the superconducting phase of the cuprate HTSC [26] with their Mott insulating behavior at half-filling, i.e. the Gutzwiller constraint of no-double-occupancy. The central hypothesis of pSO(5) theory is that this projected model is accurate in describing both the static and the dynamic properties of the HTSC. The numerical studies of two and three dimensional systems, using the SSE method, show that the pSO(5) model reproduces many salient features of the HTSC: such features as the global phase diagram with an antiferromagnetic and a superconducting phase, the doping dependence of the chemical potential, and the spin resonance peak observed in neutron scattering experiments. Furthermore, the question is raised whether the full SO(5) symmetry, which is microscopically broken by projecting out the doubly occupied states, is restored dynamically at a certain critical point of the pSO(5) model's phase diagram. Such a point is identified in the three dimensional system: a bicritical point at which the antiferromagnetically ordered phase, the superconducting phase and the paramagnetic phase meet. This scenario is very similar to the phase diagram of the classical SO(5) model for which X. Hu has performed a detailed scaling analysis of critical exponents in the vicinity of the bicritical point [30]. We therefore end this chapter with a similar scaling analysis for the pSO(5) model, which is numerically much more demanding since the numerical simulation is fully quantum mechanical in contrast to the classical one in the work of Hu. Parts of this chapter have been published in [27–29].

A detailed description and documentation of the program packages developed during this thesis is given in two appendices to this work. The first appendix describes the usage of the four program packages whose basic construction principles were discussed in chapters 2 to 4: the two ED codes for regular lattices and single unit cells, respectively, the SSE program package, and the computer algebra framework for the Dirac algebra of Second Quantization. The second appendix presents a side product created during the implementation of the SSE method: a generally applicable method for accelerating and optimizing discrete Fourier transformations in situations in which not all but only a relatively small subset of all points in the Fourier transformed space are needed. This type of problems often arises in solid state physics if certain correlation functions are measured as a function of distance r, but the quantities effectively looked for are the corresponding correlation functions at some selected momentum values $\hbar k$. This second appendix has been submitted for publication in [31].

1

Numerical simulations of strongly correlated electron systems

This chapter gives a short overview on the current state of numerical simulation techniques for strongly correlated electron systems. At first we review some frequently studied microscopic models for strongly correlated electron systems as well as the most important simulation algorithms. Then we discuss some difficulties which arise if properties of the almost infinite real system are to be inferred from simulations of small finite clusters of containing only a few dozens or a few hundreds of atoms.

1.1 Microscopic models

1.1.1 The Hubbard model

The rich and complex physics of strongly correlated electron systems is caused by the interplay of the valence electrons' kinetic energy and their potential energy due to electron-electron interaction. The Hubbard model is probably the most elementary microscopic Hamiltonian which captures this competition of kinetic and potential energy. For this reason it has been one of the most extensively studied microscopic models for correlated electron systems since its first formulation in 1963 [11]. Its Hamiltonian has the form

$$H = U \sum_{i} n_{i,\uparrow} n_{i,\downarrow} - t \sum_{\langle i,j \rangle,\sigma} (c^{\dagger}_{i,\sigma} c_{j,\sigma} + \text{h.c.}) , \tag{1.1}$$

where i and j are site indices, $\langle i,j \rangle$ are all pairs of nearest neighbor sites, $n_{i,\sigma} = c^{\dagger}_{i,\sigma} c_{i,\sigma}$ is the number of spin-σ electrons on site i, $c_{j,\sigma}$ and $c^{\dagger}_{i,\sigma}$ are electron annihilation and creation operators, and U and t are positive interaction constants.

The t-term generates hopping processes of electrons between nearest neighbor sites and thus models the kinetic energy of the system. The U-term models the Coulomb repulsion of two electrons with opposite spin in the same orbital, i.e. a potential energy. Electrostatic interaction between electrons in different orbitals is thus neglected and all orbitals

are implicitly assumed to be nondegenerate and of s-like symmetry. However, both restrictions – neglecting off-site repulsion and non s-like orbital shapes – are justified in the case of atoms with valence shells of small radius, as it occurs in the $3d$ shell of transition metals or the $4f$ shell of rare earth compounds [2, 32]. It has even been shown that the simple Hubbard model on a two-dimensional square lattice is in fact a good approximation for the low energy physics of the doped CuO_2 planes in the cuprate HTSC [33].
The interplay of potential and kinetic energy is crucial for most electronic properties of solid matter: the potential energy term is diagonal in real space and tends to immobilize the electrons at certain positions of minimum energy; the kinetic energy term, on the contrary, is diagonal in momentum space and tries to move around the electrons, which reduces the system's energy due to the Heisenberg uncertainty relation. Hence, both parts of the Hubbard model are trivial to solve and well understood, but the physics of the full Hubbard model is complex and still subject of ongoing analytical and numerical research. Several extensions to the Hubbard model are feasible and have been discussed in the literature. The most natural ones are to extend the Coulomb interaction to neighbored orbitals, i.e. to introduce an off-site potential

$$H_V = V \sum_{\langle i,j \rangle} n_i n_j$$

and/or to add hopping terms beyond nearest neighbor sites, e.g. second nearest neighbor hopping

$$H_{t'} = -t' \sum_{\langle\langle i,j \rangle\rangle,\sigma} (c^\dagger_{i,\sigma} c_{j,\sigma} + \text{h.c.}) \,,$$

where $\langle\langle i,j \rangle\rangle$ are all pairs of second nearest neighbor sites.
At half-filling, i.e. at a mean electron density ρ of exactly one electron per site, the competition between kinetic and potential energy has particularly interesting consequences: in the limit of strong on-site Coulomb repulsion ($U \gg t$), the simultaneous occupation of a site by two electrons is energetically strongly disfavored, hence each site will be occupied by exactly one electron. This means that every hopping process of a single electron costs the system the large energy U; therefore, the system is in an insulating state in which all electrons are localized. This type of insulator, in which only the pure electron-electron interaction generates the insulating behavior, is called a *Mott-Hubbard insulator*.
In the other limit, $U = 0$, the Hubbard model reduces to the *tight-binding model*

$$H = -t \sum_{\langle i,j \rangle,\sigma} c^\dagger_{i,\sigma} c_{j,\sigma} + \text{h.c.} = \sum_{\boldsymbol{k},\sigma} \varepsilon(\boldsymbol{k}) c^\dagger_{\boldsymbol{k},\sigma} c_{\boldsymbol{k},\sigma}, \tag{1.2}$$

whose band structure is given in Fig. 1.1. The figure shows that gapless excitations around $\omega = 0$ are possible at $\mathbf{k} = (\pi, 0)$ and $\mathbf{k} = (\pi/2, \pi/2)$: the system is metallic.
However, many calculations [34] indicate that on a two-dimensional lattice the ground state (i.e. the state at $T = 0$) of the Hubbard model is insulating even at infinitesimally small positive values of U. In three dimensions we expect, although no numerical studies have been performed so far, this long-range order and the insulating behavior already at finite temperatures below the Néel temperature T_N [4, 35]. This type of insulator, called *Mott-Heisenberg insulator*, is due to long-range antiferromagnetic (AF) ordering. The AF

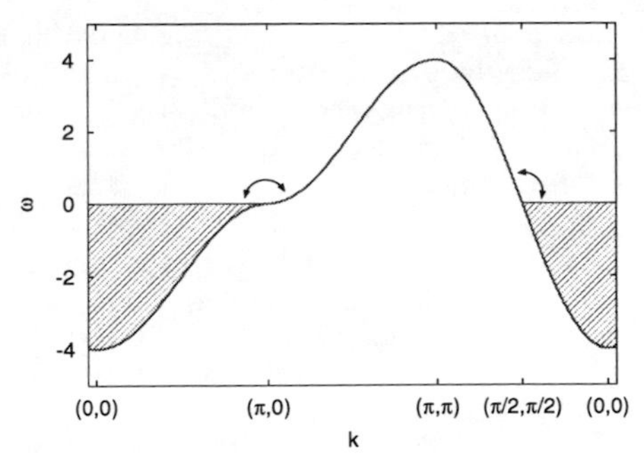

Figure 1.1: Band structure of the tight-binding model on a two-dimensional square lattice. At half-filling all the states below $\omega = 0$ are occupied; hence, gapless charge excitations are possible, indicated by the double-arrows. (Picture by M.G. Zacher)

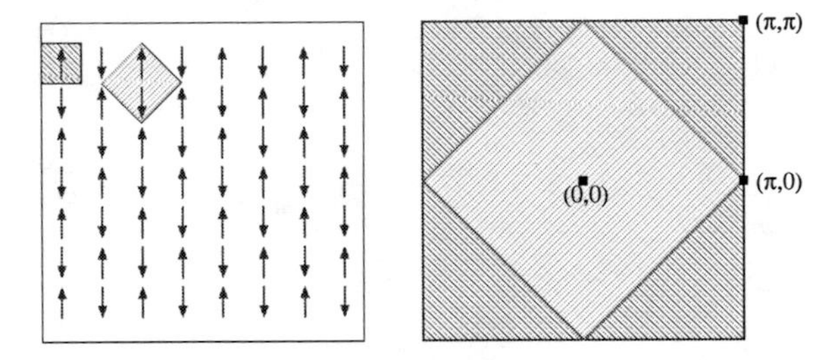

Figure 1.2: The long-range AF order effectively doubles the unit cell of the lattice in real space (left) and halves the Brillouin zone in momentum space (right). (Picture by M.G. Zacher)

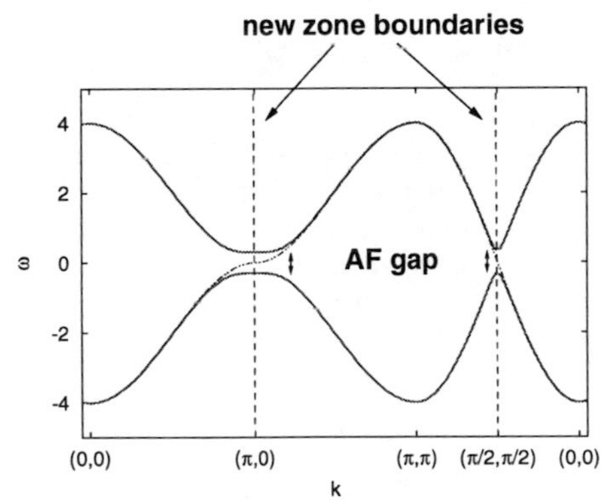

Figure 1.3: The energy bands outside the magnetic Brillouin zone are folded inside. The reflected bands at the zone boundaries $(\pi, 0)$ and $(\pi/2, \pi/2)$ trivially yield an energy gap. At half-filling, no gapless charge excitations are possible: the system is insulating. (Picture by M.G. Zacher)

configuration is energetically stabilized by the fact that neighbored spins with antiparallel alignment can reduce their energy by hopping processes; parallel spins can't do this due to the Pauli principle. The AF alignment effectively doubles the unit cell of the lattice and halves the first Brillouin zone in momentum space (see Fig. 1.2. The tight-binding band from Fig. 1.1 is reflected at the new zone boundary, and at the boundary itself Bragg reflection opens a gap [6]: the system becomes an insulator (see Fig. 1.3).
Above considerations show that the Hubbard model can be expected to have a relatively rich phase diagram with at least two different competing insulating phases an a metallic phase. This question will be studied in detail in section 2.4.1 of this work.

1.1.2 The t-J model

In the case of strong electron-electron repulsion $U \gg t$ and for electron densities of not more than one per lattice site, one may expect that states with doubly occupied sites play only a vanishingly small role in the lowest energy states of the system. The Hubbard model can then be mapped onto an effective model operating on the restricted Hilbert space in which all states with double occupancies have been projected out [36,37]. The resulting Hamiltonian is [2]

$$\begin{aligned} H = & -t \sum_{\langle i,j \rangle,\sigma} (\hat{c}^\dagger_{i,\sigma}\hat{c}_{j,\sigma} + \text{h.c.}) + J \sum_{\langle i,j \rangle} (\mathbf{S}_i \cdot \mathbf{S}_j - \frac{n_i n_j}{4}) \\ & + \frac{J}{4} \sum_{\langle i,j \rangle} \sum_{\langle j,k \rangle,\sigma} (\hat{c}^\dagger_{i,\sigma} n_{j,-\sigma} \hat{c}_{k,\sigma} - \hat{c}^\dagger_{i,-\sigma} S^\sigma_j \hat{c}_{k,\sigma}) \end{aligned} \tag{1.3}$$

with $J = 4t^2/U$ and with $\hat{c}^\dagger_{i,\sigma}$ and $\hat{c}_{i,\sigma}$ being the restrictions of $c^\dagger_{i,\sigma}$ and $c_{j,\sigma}$ onto the subspace of no-double-occupancies:

$$\hat{c}^\dagger_{i,\sigma} = c^\dagger_{i,\sigma}(1 - n_{i,-\sigma}) \quad \text{and} \quad \hat{c}_{i,\sigma} = c_{i,\sigma}(1 - n_{i,-\sigma}). \tag{1.4}$$

$\mathbf{S}_i$ is the electron spin operator on site i, and S^σ_j stands for the ladder operators S^+_j and S^-_j. The second term in (1.3) is a magnetic spin-spin exchange process which describes the kinetic exchange energy due to two 'virtual' hopping processes of electrons on neighbored sites. The third term represents two conditional hopping processes; these processes require the coordinated action of four fermion creators and annihilators (instead of two in the normal hopping processes), so it is reasonable to expect that these processes are of minor importance and can be neglected. The remaining Hamiltonian

$$H = -t \sum_{\langle i,j \rangle,\sigma} (\hat{c}^\dagger_{i,\sigma}\hat{c}_{j,\sigma} + \text{h.c.}) + J \sum_{\langle i,j \rangle} (\mathbf{S}_i \cdot \mathbf{S}_j - \frac{n_i n_j}{4}) \tag{1.5}$$

is the so-called t-J model [33,41,42] . The t-J Hamiltonian (1.5) has also been shown to be the simplest effective model for the CuO_2 planes of the cuprate HTSC [33,38], with widely accepted values of $t \approx 440$ meV and $J \approx 130$ meV [2,39,40], which corresponds to a Coulomb repulsion of $U/t \approx 12$ in the Hubbard model. More details on the physical relevance of the t-J model can be found in [41,42].

1.1.3 The Heisenberg model

At half-filling, i.e. at a mean electron density of 1 per site, the t-J model reduces to the isotropic Heisenberg model

$$H = J\sum_{\langle i,j\rangle} \mathbf{S}_i \cdot \mathbf{S}_j = J\sum_{\langle i,j\rangle} S_i^z S_j^z + \frac{J}{2}\sum_{\langle i,j\rangle} \left(S_i^+ S_j^- + S_i^- S_j^+\right). \tag{1.6}$$

The term $J\sum_{\langle i,j\rangle} n_i n_j/4$ in (1.5) only induces an overall energy offset $\Delta E(\langle n\rangle)$ and has been omitted. This model describes the magnetic interaction of localized spins; its $T=0$ ground state is antiferromagnetically ordered for $J>0$ and ferromagnetic for $J<0$. If the $S^z S^z$ part and the S^+S^- part have different interaction constants J_z and J_{xy}, the model is called *anisotropic Heisenberg model*. The anisotropic Heisenberg model with $J_z=0$ is known as the *quantum XY model*.
The antiferromagnetic spin-1/2 Heisenberg model with $J\approx 130$ meV is a simple effective model for the low energy physics of the undoped (i.e. half-filled) CuO_2 planes in the cuprates. A detailed study of other physical systems which can be described by the Heisenberg model, can be found in [43].

1.1.4 Effective bosonic models for the Heisenberg model

The low energy excitations of correlated electrons systems can often be described in terms of bosonic quasiparticles which are created out of the system's ground state. The ground state can then be considered as an effective 'vacuum' state, and the entire physics of the system is captured by only modeling the bosonic excitations. This mapping of the fermionic physical system onto a bosonic effective system has several advantages. First, this approach helps to work out clearly which processes and which excitations dominate the low energy physics of the system. Second, bosonic systems are often much easier to analyse numerically as one does not have to bother about Pauli principles, Slater determinants, and Fermi signs. Furthermore, when performing QMC simulations, the infamous sign problem (see section 1.2.3) is less likely to occur in bosonic systems – at least if there are no inherent frustrations in the system.
We want to demonstrate the procedure at the example of the two-dimensional (2D) Heisenberg model. This model (with $J > 0$) is the simplest effective model describing the cuprates' CuO_2 planes near zero doping, i.e. in the range in which the cuprates have an antiferromagnetically (AF) ordered phase at low temperature.[1] Since we want to construct *bosonic* quasiparticles, we have to divide the lattice into effective sites containing an *even* number of fermionic sites (with one electron per site). In order to conserve the symmetry between x and y-direction in the system, we choose a plaquette of 2×2 fermionic sites. The low energy eigenstates of this Heisenberg plaquette are determined very easily using the computer algebra program described in chapter 4. We find the nondegenerate

[1]The Mermin-Wagner theorem [35] excludes the existence of a phase with real long-range AF order in two dimensions at $T > 0$; however, we can still expect a good modeling of the paramagnetic high temperature phase, and at low temperatures at least a quasi long-range AF order is possible.

ground state

$$|\,0\,\rangle = \frac{1}{2\sqrt{3}}\Big(2\left|\begin{smallmatrix}\uparrow & \downarrow \\ \downarrow & \uparrow\end{smallmatrix}\right\rangle + 2\left|\begin{smallmatrix}\downarrow & \uparrow \\ \uparrow & \downarrow\end{smallmatrix}\right\rangle - \left|\begin{smallmatrix}\uparrow & \uparrow \\ \downarrow & \downarrow\end{smallmatrix}\right\rangle - \left|\begin{smallmatrix}\downarrow & \downarrow \\ \uparrow & \uparrow\end{smallmatrix}\right\rangle - \left|\begin{smallmatrix}\uparrow & \downarrow \\ \uparrow & \downarrow\end{smallmatrix}\right\rangle - \left|\begin{smallmatrix}\downarrow & \uparrow \\ \downarrow & \uparrow\end{smallmatrix}\right\rangle\Big)$$

with energy $E_0 = -2J$ and total spin $S = 0$. This singlet state will be the vacuum state $|\,vac\,\rangle$ of our effective bosonic model. The next energy eigenstates are three triplet states $|\,t_i\,\rangle$ with energy $E_t = -J$ and spin quantum numbers of $S = 1$ and $S^z = -1$, 0 and 1:

$$\begin{aligned}
|\,t_-\rangle &= \frac{1}{2}\ \Big(\left|\begin{smallmatrix}\downarrow & \downarrow \\ \downarrow & \uparrow\end{smallmatrix}\right\rangle - \left|\begin{smallmatrix}\downarrow & \downarrow \\ \uparrow & \downarrow\end{smallmatrix}\right\rangle + \left|\begin{smallmatrix}\uparrow & \downarrow \\ \downarrow & \downarrow\end{smallmatrix}\right\rangle - \left|\begin{smallmatrix}\downarrow & \uparrow \\ \downarrow & \downarrow\end{smallmatrix}\right\rangle\Big), \\
|\,t_z\rangle &= \frac{1}{\sqrt{2}}\Big(\left|\begin{smallmatrix}\uparrow & \downarrow \\ \downarrow & \uparrow\end{smallmatrix}\right\rangle - \left|\begin{smallmatrix}\downarrow & \uparrow \\ \uparrow & \downarrow\end{smallmatrix}\right\rangle\Big), \\
|\,t_+\rangle &= \frac{1}{2}\ \Big(\left|\begin{smallmatrix}\uparrow & \uparrow \\ \uparrow & \downarrow\end{smallmatrix}\right\rangle - \left|\begin{smallmatrix}\uparrow & \uparrow \\ \downarrow & \uparrow\end{smallmatrix}\right\rangle + \left|\begin{smallmatrix}\downarrow & \uparrow \\ \uparrow & \uparrow\end{smallmatrix}\right\rangle - \left|\begin{smallmatrix}\uparrow & \downarrow \\ \uparrow & \uparrow\end{smallmatrix}\right\rangle\Big).
\end{aligned}$$

In order to treat the spatial dimensions x, y and z symmetrically, it is convenient to transform $|\,t_-\rangle$ and $|\,t_+\rangle$ into two new triplet states

$$\begin{aligned}
|\,t_x\rangle &= \frac{1}{\sqrt{2}}(|\,t_-\rangle + |\,t_+\rangle) \\
|\,t_y\rangle &= \frac{\mathrm{i}}{\sqrt{2}}(|\,t_-\rangle - |\,t_+\rangle)\,.
\end{aligned}$$

The three states $|\,t_x\,\rangle$, $|\,t_y\,\rangle$ and $|\,t_z\,\rangle$ can now be identified with three bosonic quasiparticles t_x, t_y and t_z representing the lowest energy excitations of the Heisenberg plaquette:

$$|\,t_{x,y,z}\rangle = t^\dagger_{x,y,z}\,|\,vac\,\rangle\ .$$

All other energy eigenstates of the Heisenberg plaquette have energies $E \geq 0$ and can be neglected in the low energy effective model. It should be noted that the quasiparticles t_x, t_y and t_z, which carry spin 1 and charge 0, are *hardcore bosons* because one cannot create more than one of them simultaneously on a single plaquette.
After having identified the particles in our effective model, the next task is to construct an effective Hamiltonian. This Hamiltonian must fulfill two requirements. First, it should contain a chemical-potential-like term modeling the excitation energy of $\Delta E = -J - (-2J) = J$ which is needed to create a triplet excitation on a plaquette in the singlet (vacuum) state:

$$H_1 = J \sum_{r,\,\alpha=x,y,z} t^\dagger_{\alpha(r)}\, t_{\alpha(r)}\,, \tag{1.7}$$

where r runs over all plaquettes of the system and α labels the three particle types. Second, the Hamiltonian should contain a term modeling the interaction between nearest neighbor plaquettes. Because of the hardcore constraint, this second part of the Hamiltonian must be of the general form

$$H_2 = \sum_{\langle r,s\rangle,\,\alpha,\beta} \left(A_{\alpha\beta}\, t^\dagger_{\alpha(r)} t^\dagger_{\beta(s)} + B_{\alpha\beta}\, t^\dagger_{\alpha(r)} t_{\beta(s)} + C_{\alpha\beta}\, t_{\alpha(r)} t^\dagger_{\beta(s)} + D_{\alpha\beta}\, t_{\alpha(r)} t_{\beta(s)}\right), \tag{1.8}$$

where $\langle r, s\rangle$ are all pairs of nearest neighbor plaquettes and $\alpha, \beta = x, y, z$ stand for the particle types or triplet components. The weight prefactors $A_{\alpha\beta}$ to $D_{\alpha\beta}$ can be related to the value of J in the microscopic Heisenberg model by comparing the interaction processes between two adjacent plaquettes in the original microscopic description and in the effective plaquette description of the system (see Fig. 1.4). In the 'microscopic' description, the

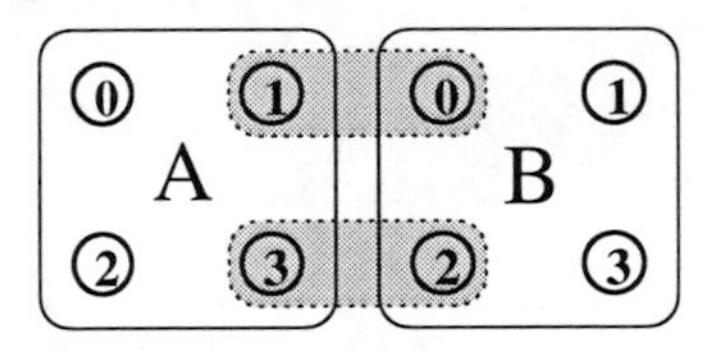

Figure 1.4: Two adjacent plaquettes A and B of the bosonic effective model interact via two 'microscopic' electron-electron interactions (between sites 1A and 0B, and 3A and 2B) of the original Heisenberg model.

interaction Hamiltonian between the plaquettes A and B has the form

$$H_{AB}^z = J\big(S_{1A}^z S_{0B}^z + S_{3A}^z S_{2B}^z\big). \tag{1.9}$$

(We only have given the z part of the interaction, but due to the SO(3) rotational invariance of the Heisenberg Hamiltonian (1.6) the x and y components can be treated analogously.) In the effective bosonic description, the interaction term H_{AB}^z creates the terms of H_2 in (1.8). We therefore have

$$\begin{aligned}\big\langle t_z(A)\, t_z(B)\,\big|H_{AB}^z\big|\,vac(A)\, vac(B)\big\rangle &= A_{zz}\,,\\ \big\langle t_z(A)\, vac(B)\,\big|H_{AB}^z\big|\,vac(A)\, t_z(B)\big\rangle &= B_{zz}\,,\end{aligned} \tag{1.10}$$

and analogous expressions can be given for all interaction processes in (1.8). The two equations (1.10) are easily solved by means of the computer algebra package described in chapter 4 of this work. The result is

$$\begin{aligned}A_{zz} &= J\Big(\big\langle t_z\big|S_1^z\big|vac\big\rangle_{(A)}\big\langle t_z\big|S_0^z\big|vac\big\rangle_{(B)} + \big\langle t_z\big|S_3^z\big|vac\big\rangle_{(A)}\big\langle t_z\big|S_2^z\big|vac\big\rangle_{(B)}\Big) = -\frac{J}{3}\\ B_{zz} &= J\Big(\big\langle t_z\big|S_1^z\big|vac\big\rangle_{(A)}\big\langle vac\big|S_0^z\big|t_z\big\rangle_{(B)} + \big\langle t_z\big|S_3^z\big|vac\big\rangle_{(A)}\big\langle vac\big|S_2^z\big|t_z\big\rangle_{(B)}\Big) = -\frac{J}{3}\,,\end{aligned}$$

and analogously $C_{zz} = -J/3$, $D_{zz} = -J/3$. The same result is obtained for the two other spatial directions x and y. Similarly, one proves that all mixed terms $X_{\alpha\beta}$ (with $X = A, B, C, D$ and $\alpha \neq \beta$) vanish.

In summary, we have determined the following effective bosonic model for the 2D Heisenberg model:

$$H_{eff} = J \sum_{r,\,\alpha=x,y,z} t_{\alpha}^{\dagger}(r) t_{\alpha}(r) - \frac{J}{3} \sum_{\langle r,s\rangle\,\alpha=x,y,z} \big(t_{\alpha}^{\dagger}(r) + t_{\alpha}(r)\big)\big(t_{\alpha}^{\dagger}(s) + t_{\alpha}(s)\big)\,, \tag{1.11}$$

where t_x, t_y and t_z are three hardcore boson quasiparticles corresponding to the lowest energy spin excitations on the 2×2 Heisenberg plaquette.

For the AF Heisenberg model on the quasi one-dimensional (1D) system of a two-leg ladder, a bosonic description by means of bosonic spin excitations is even more straight forward than in the 2D case. On the ladder, the rungs form natural effective 'plaquettes'. An SO(3)-symmetric effective hardcore boson Hamiltonian for the ladder has been derived in [44, 45]; its Hamiltonian reads

$$
\begin{aligned}
H \;=\; & J\sum_n \mathbf{t}_n^\dagger\cdot\mathbf{t}_n + \frac{J}{2}\sum_n(\mathbf{t}_n^\dagger\cdot\mathbf{t}_{n+1}^\dagger + \text{h.c.}) + \frac{J}{2}\sum_n(\mathbf{t}_n^\dagger\cdot\mathbf{t}_{n+1} + \text{h.c.}) \\
& - \frac{J}{4}\sum_n(1-\delta_{\alpha\beta})(\mathbf{t}_n^\dagger\cdot\mathbf{t}_{n+1}^\dagger\,\mathbf{t}_{n+1}\cdot\mathbf{t}_n - \mathbf{t}_n^\dagger\cdot\mathbf{t}_{n+1}\,\mathbf{t}_{n+1}^\dagger\cdot\mathbf{t}_n)
\end{aligned}
\tag{1.12}
$$

Here, $\mathbf{t}$ is the vector containing the three triplet bosons: $\mathbf{t} = (t_x, t_y, t_z)^T$. This Hamiltonian forms the basis for the bosonic description of the Heisenberg ladder with nonmagnetic impurities which will be discussed in section 3.3.3 of this thesis.

1.1.5 Four-boson models

In the previous section we have constructed an effective bosonic description for the 'AF regime' of cuprate HTSC near zero doping. Now we want to enlarge this description by adding a fourth hardcore boson carrying spin 0 and charge 2 which describes the formation of Cooper pairs in the hole doped 'SC regime' of the cuprates.
Our starting point is again the 2x2 plaquette of the two-dimensional AF Heisenberg model, but this time we study the low energy physics of plaquettes carrying one or two holes. The one-hole ground state, i.e. the ground state of the 2×2 plaquette with 3 electrons, is a fourfold degenerate spin-1/2 state with energy $E=-J$; the two-hole ground state turns out to be also fourfold degenerate, with spin $S=0$ and energy $E=-3J/4$. Hence, the 2D Heisenberg model can explain why the holes in the cuprates stick together and form Cooper pairs: the minimum total energy of a half filled and a two-hole plaquette, $E_{tot}=-2.75\,J$, is considerably lower than twice the ground state energy of the one-hole plaquette.
This statement remains valid if we study the physically more realistic t-J model (1.5), in which the holes are allowed to move to nearest neighbor sites. Another effect of switching on the hopping is to remove the degeneracy of the ground state in both the one-hole and the two-hole system. In the two-hole case, the ground state turns out to be a linear combination of

$$
\left|\begin{smallmatrix}\uparrow & - \\ - & \downarrow\end{smallmatrix}\right\rangle - \left|\begin{smallmatrix}\downarrow & - \\ - & \uparrow\end{smallmatrix}\right\rangle + \left|\begin{smallmatrix}- & \uparrow \\ \downarrow & -\end{smallmatrix}\right\rangle - \left|\begin{smallmatrix}- & \downarrow \\ \uparrow & -\end{smallmatrix}\right\rangle
$$

and

$$
\left|\begin{smallmatrix}\uparrow & \downarrow \\ - & -\end{smallmatrix}\right\rangle - \left|\begin{smallmatrix}\downarrow & \uparrow \\ - & -\end{smallmatrix}\right\rangle + \left|\begin{smallmatrix}- & - \\ \uparrow & \downarrow\end{smallmatrix}\right\rangle - \left|\begin{smallmatrix}- & - \\ \downarrow & \uparrow\end{smallmatrix}\right\rangle + \left|\begin{smallmatrix}- & \uparrow \\ - & \downarrow\end{smallmatrix}\right\rangle - \left|\begin{smallmatrix}- & \downarrow \\ - & \uparrow\end{smallmatrix}\right\rangle + \left|\begin{smallmatrix}\uparrow & - \\ \downarrow & -\end{smallmatrix}\right\rangle - \left|\begin{smallmatrix}\downarrow & - \\ \uparrow & -\end{smallmatrix}\right\rangle,
$$

in which the relative weight of both contributions depends on the ratio t/J. We call this non-degenerate singlet ground state $|\,t_h\,\rangle$ and enlarge our model (1.11) by a fourth hardcore boson t_h defined by

$$
|\,t_h\rangle = t_h^\dagger\,|\,vac\rangle\ .
$$

When constructing the enlarged Hamiltonian, we have to keep in mind that the new operators $t_h^\dagger$ and t_h change the electron number of the system by ± 2. Therefore, the chemical-potential-like term for the new boson – corresponding to (1.7) – has to incorporate the real chemical potential μ in the form

$$H_{1,c} = (\Delta_c - 2\mu) \sum_r t_{h(r)}^\dagger t_{h(r)} \,. \tag{1.13}$$

Here, we do not try to express the new constant Δ_c, which describes the energy needed to create a hole pair boson, in terms of the microscopic parameters t and J. We mearly note that the value of Δ_c depends on the ratio t/J and has the only effect to redefine the zero of the chemical potential.
In the interaction part – corresponding to (1.8) – we omit the pair creation and annihilation terms $t^\dagger t^\dagger$ and $t\,t$ since the overall number of charged particles in the system cannot change. We therefore write

$$H_{2,c} = -J_c \sum_{\langle r,s\rangle} \left(t_{h(r)}^\dagger t_{h(s)} + \text{h.c.} \right) . \tag{1.14}$$

Again, the value of the new parameter J_c depends on t/J; we consider J_c a free parameter of the four-boson model and do not try to express it in terms of the 'microscopic' parameters t and J. [2]
The general form of the four-boson model finally reads

$$\begin{aligned} H \;=\; & J \sum_{r,\alpha=x,y,z} t_{\alpha(r)}^\dagger t_{\alpha(r)} + (\Delta_c - 2\mu) \sum_r t_{h(r)}^\dagger t_{h(r)} \\ & - \frac{J}{3} \sum_{\langle r,s\rangle\, \alpha=x,y,z} \left(t_{\alpha(r)}^\dagger + t_{\alpha(r)} \right) \left(t_{\alpha(s)}^\dagger + t_{\alpha(s)} \right) - J_c \sum_{\langle r,s\rangle} \left(t_{h(r)}^\dagger t_{h(s)} + \text{h.c.} \right). \end{aligned} \tag{1.15}$$

Interestingly, a Hamiltonian of this form can also be derived from SO(5) theory. We will study this so-called projected SO(5) model in detail in chapter 5.

1.2 Numerical simulation techniques

During the last decade, several algorithms and simulation techniques have been developed with which finite systems – called 'clusters' – carrying about 10 to 1000 particles can be studied numerically. The three most important techniques are Exact Diagonalization (ED) [1, 2, 48], Density Matrix Renormalization Group (DMRG) methods [49 51] and Quantum Monte Carlo simulations (QMC) [18, 52–58]. However, the correct extrapolation from the finite cluster results to the infinite system is a nontrivial task and several caveats have to be taken in mind (see section 1.3). For this reason, there have been several approaches to incorporate the transition to the infinite lattice into a 'second layer' within the numerical simulation. One important such approach is the Cluster Expansion Technique (CET) [63].

[2] Such a mapping between microscopic fermionic and effective bosonic interaction parameters for various microscopic models has been performed by Altman *et al.* [46] by means of renormalization group methods.

1.2.1 Exact Diagonalization

Systems of quantum mechanical particles on the discrete points of a lattice, e.g. electrons in a crystal lattice, can be described by a finite-dimensional linear space, called Hilbert space; physically measurable quantities become linear operators in this space, and after the choice of a basis they adopt the form of finite-dimensional matrices. The physical properties of the studied system are then expressed by the eigenvalues and eigenvectors of these matrices, which can be obtained from diagonalizing the matrices. This approach is called "Exact Diagonalization". It has the advantage to provide an exact access not only to ground state properties of the system but also to its dynamics, i.e. its reaction to external perturbations. On the other hand, this method is limited to very small system sizes of typically 16 to 40 lattice points and particles since the Hilbert space dimension grows exponentially with system size and the number of particles. The ED technique will be described in detail in chapter 2 of this thesis.

1.2.2 Density Matrix Renormalization Group

This approach is essentially a combination of Exact Diagonalization with renormalization group techniques. Subsystems of different length scales are diagonalized exactly, and in each step from a smaller to a larger subsystem the exponential growth of Hilbert space is avoided by only keeping those base states which contribute most to the low-energy physical properties of the whole system. Thus, the practically accessible system size is increased by a factor of 2 to 5 with respect to ED. However, the method works best for low dimensional systems (e.g. 1D chain lattices) and rapidly looses performance when going to 2D or even 3D systems. A review on this technique can be found in [59].

1.2.3 Quantum Monte Carlo simulations

QMC techniques are currently the most important simulation techniques of theoretical solid state physics. Many different variants have been developed and we refer to [18] for an overview. The basic idea of all these variants is not to use *all* basis states of Hilbert space for exactly diagonalizing the measurement operators, but to calculate statistical mean values of the measurable quantity based on a relatively small ensemble of the 'statistically most relevant' states of the system. These states are sampled starting from an arbitrarily chosen initial state and using a statistical sampling process. The statistical process has to be organized in such a way that the correct thermodynamics of the studied system is captured, i.e. each state s should be sampled with a probability which is proportional to its Boltzmann factor $\mathrm{e}^{-E(s)/k_BT}$ [60, 61]. It can be shown [60] that this is guaranteed if the process fulfills the *detailed balance* criterion which states that for each transition probability $\mathcal{P}$ from a state s_1 to another state s_2 we must have

$$\mathcal{P}(s_1 \rightarrow s_2)\mathrm{e}^{-E(s_1)/k_BT} = \mathcal{P}(s_2 \rightarrow s_1)\mathrm{e}^{-E(s_2)/k_BT}\,. \tag{1.16}$$

Here, $E(s_1)$ and $E(s_2)$ are the energies of s_1 and s_2. When constructing a statistical process, the detailed balance criterion has to be cast into an acceptance probability rule. Such a rule decides whether a new state s_2 constructed from the current state s_1 is accepted or rejected. One easily verifies that the *Metropolis* [62] acceptance probability

$$p(s_1 \to s_2) = \begin{cases} e^{-(E(s_1)-E(s_2))/k_B T} & \text{if} \quad E(s_2) > E(s_1) \\ 1 & \text{if} \quad E(s_2) < E(s_1) \end{cases} \tag{1.17}$$

is suitable for this task. The Metropolis rule always accepts a new state if its energy is lower than the current state's energy. Otherwise, the new state is only accepted with a probability $p<1$ which exponentially decreases to zero as the energy difference increases. The QMC approach allows to treat systems of several thousand lattice sites and particles, and unlike ED and DMRG (apart from some rather involved extensions to $T>0$), it also works at finite temperature, thus providing access to thermodynamical properties of the system. However, when treating fermionic systems (like electrons) or systems with inherent frustrations, one often runs into the infamous minus sign problem [18], which severely reduces the accessible system size and temperature range.

1.2.4 Cluster Expansion Techniques

The simulation techniques presented above often suffer from the limited system size they are able to deal with. For this reason, a new approach allowing to treat systems with an essentially infinite number of particles would be highly desirable. Since macroscopic many-particle systems consist of about 10^{23} particles, the 'infinite-system-approach' is much more suitable and accurate in describing these systems' physics than any numerical study of several tens or hundreds of particles could ever be.
Cluster Expansion Technique (CET) is such an approach [63]; its basic idea is to connect infinitely many finite clusters – whose physics can be extracted using any of the techniques described above – by some inter-cluster interaction treated as a perturbation expansion. More precisely, let the extracted property of the cluster be contained in G_c, the so-called Green's function of the cluster. So far the particles move and interact within a given cluster only. How are they proliferating out of the cluster, thereby forming an essentially infinite electronic system (such as in the CuO_2 plane of the HTSC)? In the simplest version of the Cluster Expansion Technique the proliferation proceeds iteratively, in the sense of a geometrical series, i.e.

$$\begin{aligned} G_\infty &= G_c + G_c T G_c + G_c T G_c T G_c + \ldots \\ &= \frac{G_c}{I - T G_c}. \end{aligned} \tag{1.18}$$

Here, G_∞ stands for the corresponding Green's function containing the electronic properties of the infinite lattice. T is an $M \cdot M$ matrix, with M being the cluster size. It describes the proliferation out of one cluster into the neighboring one. The geometrical series in eq. (1.18), then, corresponds to an iterated proliferation all over the infinite-size lattice.
This, in principle, rather straightforward idea yields amazingly accurate results: it can be shown to be exact in two extremely different albeit special limits: in the limit of vanishing interaction between the particles and in the opposite limit of a vanishing ratio between kinetic energy and electric Coulomb repulsion among particles. It is clear that eq. (1.18) is opening up a new route towards the "thermodynamic limit", i.e. the infinite-size limit,

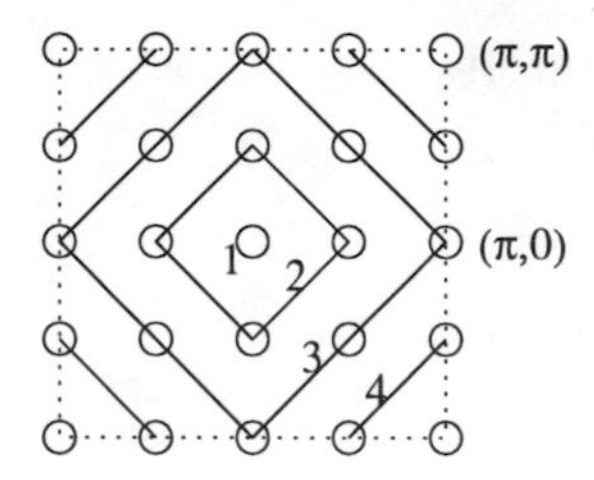

Figure 1.5: The noninteracting tight-binding Hamiltonian $H = -t\sum_{\langle i,j\rangle,\sigma}(c^\dagger_{i,\sigma}c_{j,\sigma} + h.c.)$ is diagonal in **k***-space with identical energy eigenvalues on each of the shells labelled 1, 2, 3 and 4 in above figure; the eigenenergy increases with the shell number. The open circles are the allowed* **k** *points of the 4×4 cluster. For all electron numbers besides 2, 10, 22, 30 and 32 there are partially filled shells, and the ground state is (highly) degenerate.*

which is of fundamental interest for a large variety of solid-state phenomena, for example phase transitions. Just to mention an example for such a fundamental aspect in connection with the HTSC materials, consider the nature of the superconducting state itself. This state is neccessarily a "macroscopic" state, describing the coherent propagation of about 10^{23} electrons (or, more precisely, paired electrons, the so-called Cooper pairs) through a metal, for example a superconducting wire. In order to describe this state one obviously has to go to the macroscopic limit.

1.3 Finite-size effects

1.3.1 Classes of finite-size effects

It has been mentioned in the introduction of the previous section that inferring the properties of a real physical system containing an (almost) infinite number of particles from the properties of a small finite cluster is a highly nontrivial task. Several so-called finite-size effects have to be taken into account which can cause the finite cluster to have completely different physical properties than the infinite system. Eder [2] distinguishes between five kinds of finite-size effects:

- **Shell effects and multiplet splitting**:
  The allowed momenta of a small cluster form a coarse grid of points (see eq. (2.36) in the next chapter). Due to the point group symmetry of the finite cluster the noninteracting part of many microscopic models – e.g. the hopping part of the Hubbard or t-J model – then creates shells of degenerate momenta [2] (see Fig. 1.5). Electron-electron interaction lifts the n-fold degeneracy of the ground state and creates a multiplet splitting and a multiplet structure which is completely a finite-size artefact. The numerically detected ground state then reflects many properties of the ground state multiplet and might not be a good approximation for the thermodynamical limit. In particular, the Fermi surfaces of the finite and the infinite lattice can look very different [2].

- **Broken symmetries due to quasi long-range correlations:**
Finite but quasi long-ranged static correlations induce a broken symmetry of the ground state on the finite cluster if the correlation length becomes comparable to the system size. For example, short-range static AF correlations create a finite staggered magnetization on the finite lattice, whereas this quantity vanishes on the infinite lattice.

- **Hidden symmetries of the Hamiltonian:**
It is not only possible that the infinite system shows a symmetry which the finite system does not show, the inverse situation may also appear. The ground state of the two-dimensional AF Heisenberg model for example, has spontaneously broken symmetry: the direction of the Néel vector lifts the SO(3) rotational symmetry of the system. In the finite system, on the contrary, the total spin is still a good quantum number, and all energy eigenstates are also eigenstates of $\mathbf{S}^2$.

- **Finite-size gaps:**
Gaps in the single particle spectral function or in the spin correlation function of a finite system "are the rule rather than the exception because one is always comparing systems in which the densities differ by a finite amount" [2]. Therefore, a finite gap in a small cluster's single particle spectral function (which traces the system's band structure) is in general not an unambiguous criterion for insulating behavior.

- **Lifetime effects:**
Diagonalizing small systems returns exact many-body eigenstates of infinite lifetime, and the spectral functions are a series of delta peaks. For this reason it is in general not possible to deduce peak widths, damping rates and lifetimes from numerical cluster simulations.

1.3.2 Finite-size scaling

Apart from the 'fundamental' finite-size effects discussed in the preceding section one can expect that almost all physical quantities measured on the finite cluster contain some 'smooth' modifications compared to the same properties on the infinite lattice. These modifications, which may be expected to vanish systematically with increasing system size, can be eliminated by means of a so-called finite-size scaling. The basic idea is to measure one physical observable on different cluster sizes and to plot the recorded data as a function of inverse cluster size $1/L$. Then an interpolating fit for the data points is determined, whose extrapolation to $1/L \to 0$ is (hopefully) a good approximation for the desired quantity on the infinite lattice.
As an example we consider a spatial static correlation function $\langle\langle X(r)\, X(0)\rangle\rangle$ which decays to zero as the distance r goes to infinity (see Fig. 1.6). When measured on a finite lattice, say a one dimensional chain of 8 sites, all r-points larger than 7 or smaller than 0 are mapped onto the range 0...7. Hence, in each of the 8 r-points, the function values of Unfortunately, a very exact determination of the decay behavior of correlation functions is often crucial, for example to find the transition line of a *Kosterlitz-Thouless* phase transition.[3] For such

[3] At a Kosterlitz-Thouless transition point, a correlation function $C(r)$ changes from fast exponential decay $X(r) \propto C\mathrm{e}^{-\lambda r}$ to slower power law decay $X(r) \propto Cr^{-\alpha}$.

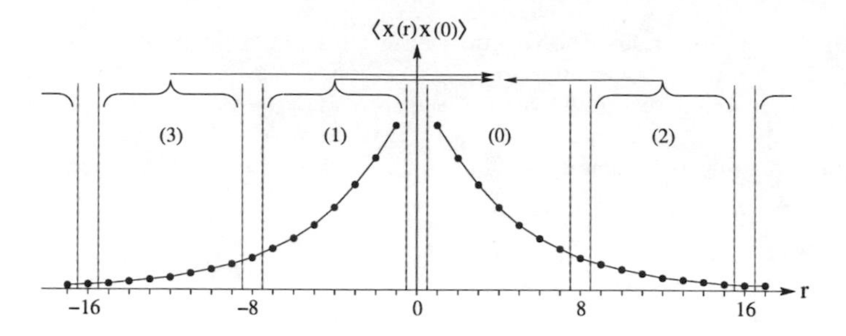

Figure 1.6: A static correlation function $\langle X(r)\,X(0)\rangle$ plotted as a function of distance r. The correlation decays to zero for $r\to\infty$. When recorded on a finite chain lattice of length 8 with periodic boundary conditions, all intervals labeled 1, 2 ,3 ... are in fact repetitions of the same 8 lattice sites.

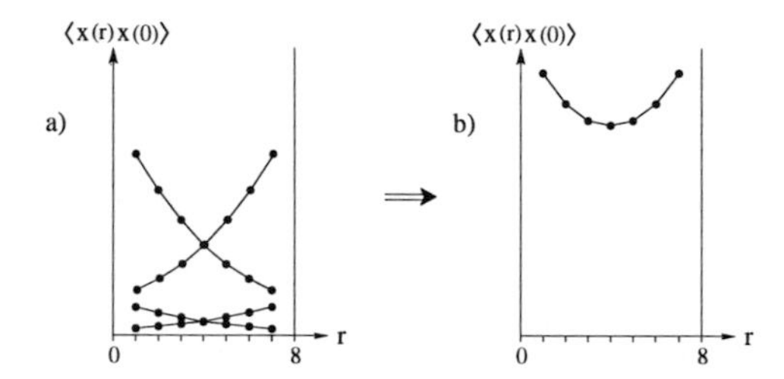

Figure 1.7: When recorded on an 8-site chain, all function values from the intervals 1, 2, 3, ... in Fig. 1.6 are superimposed (left) and contribute to the measured function values on the 8 sites of the system (right).

a subtle distinction the heavily deformed raw data in Fig. 1.7 (right) are not suitable. Here, a finite-size scaling often yields surprisingly good results: after performing finite-cluster calculations for a couple of different cluster sizes L, one traces the results as a function of $1/L$ and determines an extrapolated value for the point $1/L=0$ (see Fig. 1.8). Figure 1.9 shows the result of such a delicate determination of a Kosterlitz-Thouless transition line (KT) by means of a finite-size scaling analysis. We have measured the static spatial correlation function between hole pairs in the 2D projected SO(5) model (see chapter 5) on four different square lattices with lengths $L=12$, 14, 18 and 22 at two slightly different values of the chemical potential, $\mu=-0.05$ and $\mu=0.15$. We expect the KT transition to occur somewhere between these two values. An exponential decay of $X(r)$ yields a straight line in a $\log(X)$ versus r plot, whereas a power law decay forms a straight line in the $\log(X)$ versus $\log(r)$ plot. Figure. 1.9 compares these two types of plots. We observe a perfect power law decay at $\mu=0.15$ and an equally perfect exponential one at $\mu=-0.05$, which indeed proves the existence of a KT transition between these two values of the chemical potential.

The approach to finite-size scaling presented here is a more or less 'intuitive' one; it is legitimated by its success in practical applications rather than by a deep theoretical

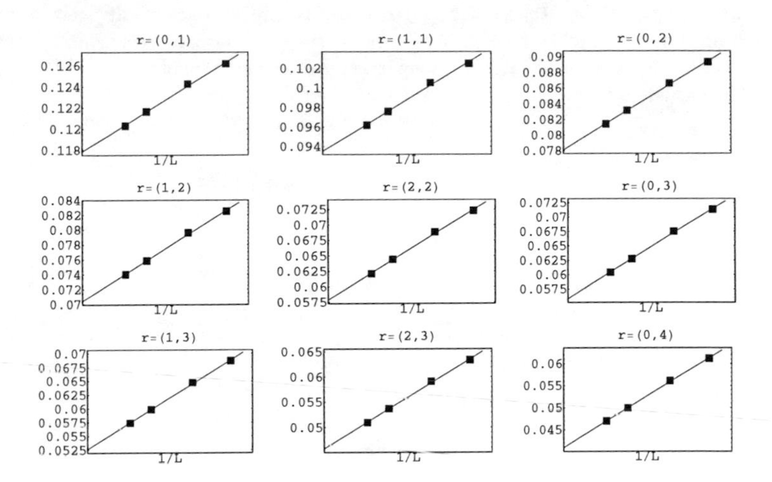

Figure 1.8: hole pair correlation function for the 2D projected SO(5) model at $T/J = 0.5$ and $\mu = -0.05$ (see chapter 5) versus inverse system size $1/L$. Each plot shows the correlation for a certain distance r. The four data points in each plot are the finite cluster measurements on square lattices of lengths $L = 12$, 14, 18 and 22.

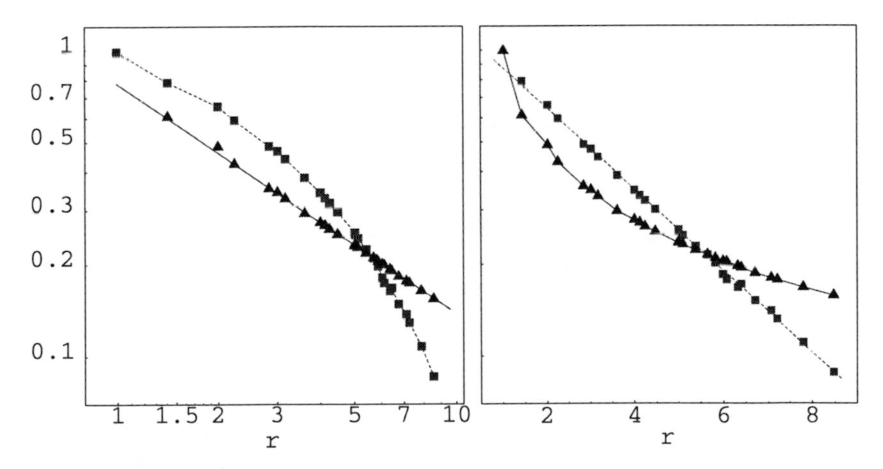

Figure 1.9: Decay behavior of the SC correlation function $X(r) = \left(t_h^\dagger(r)\, t_h(0) + \text{h.c.}\right)$ for the chemical potentials $\mu = -0.05$ (squares) and $\mu = 0.15$ (triangles) in the projected SO(5) model at temperature $T = J/2$. Each point $X(r)$ is the result of a finite-size scaling as it is shown in Fig. 1.8. The left plot is a log-log plot, the right one a simple logarithmic plot.

foundation. In particular, the choice to trace the measured data points versus inverse cluster length $1/L$ is not at all the only way to prepare the extrapolation. A tracing versus inverse cluster volume $1/L^d$ with arbitrary positive exponents d would be possible as well. For a better founded and more sophisticated finite-size scaling theory we refer to Ref. [47]

2

Exact Diagonalization

In its first part, this chapter will be devoted to a general overview on the concept of Exact Diagonalization (ED) and a concise description of the numerical Lanczos algorithm [48], followed by further sections summarizing implementation tricks and strategies for different system geometries and Hamiltonians. In the second part, two exemplary applications to current problems in solid state physics will be presented: first a detailed study of the phase diagram and the metal-insulator transition in the Hubbard model, then the simulation and interpretation of X-ray dichroism spectra of a technologically interesting magnetic metal-oxide compound, CrO_2. Parts of these works have been bublished in [12] and [14].

2.1 Basic idea

On of the most elementary tasks in theoretical solid state physics is to calculate the energy eigenvalues and eigenstates of a given model Hamiltonian. Typically, this Hamiltonian has been constructed starting from certain assumptions on the geometry and the relevant physical properties of the studied system.

If the total number of allowed states of the system (i.e. the dimension of the system's Hilbert space) is finite, this task can be solved using the tools of linear algebra: first one chooses a basis in which the Hamiltonian can be formulated easily; the resulting Hamiltonian matrix can then be diagonalized numerically or algebraically. This method is called 'exact diagonalization' (ED) because it solves the quantum mechanical eigenvalue problem exactly, without the approximations and systematic or statistical errors inevitably present in other methods like quantum Monte Carlo techniques (QMC) [52, 53, 55] or density matrix renormalization (DMRG). [49, 50]

In practice, the range of system sizes which are accessible via ED is severely limited by the exponential increase of Hilbert space dimension with the number of orbitals in the system. Let us, for example, consider a system of $n = n_\uparrow + n_\downarrow$ electrons in N_o electronic orbitals. For each orbital there are 4 allowed states, $|-\rangle, |\uparrow\rangle, |\downarrow\rangle$ and $|\uparrow\downarrow\rangle$, as the Pauli principle enforces that any two electrons differ in at least one quantum number. Hence, the dimension *dim* of Hilbert space is

n	free	fixed	fixed
$n_\uparrow$, $n_\downarrow$	free	free	fixed
dim	4^{N_o} $= 4{,}294{,}967{,}296$ for $N_o\!=\!16$	$\binom{2N_o}{n}$ $= 601{,}080{,}390$ for $N_o\!=\!16$, $n\!=\!16$	$\binom{N_o}{n_\uparrow}\binom{N_o}{n_\downarrow}$ $= 165{,}636{,}900$ for $N_o\!=\!16$, $n_\uparrow\!=\!n_\downarrow\!=\!8$

From these data it is clearly visible that a complete and exact solution of the eigenvalue problem is impossible with currently available methods and computers for systems with more than 10 or 12 orbitals. Two tricks help to push this limit to 20 or 22 orbitals: first, spatial symmetries of the system's geometry – translational, rotational and inversion symmetries – can be used to split up the full Hamiltonian into a series of smaller blocks which can be treated independently [1, 64, 65]. Second, instead of calculating *all* eigenvalues and eigenstates exactly one can refer to much faster iterative methods returning good approximations for the lowest eigenvalues only. Since the latter determine the physical properties of the system – the ground state and the low-energy excitations – the information gained in this way is in most cases sufficient to describe the system. The practically most important of these iterative algorithms is the so-called Lanczos method. [1, 48] In the literature, the expression 'exact diagonalization' is frequently used as synonym for 'approximative solution of the physical many-body eigenvalue problem by means of the Lanczos algorithm'. We follow this tradition, keeping in mind that in fact the method is not at all 'exact' but only an approximative and partial solution of the eigenvalue problem. However, since the errors in energies and energy eigenstates can easily be controlled and pushed to almost arbitrarily small values within Lanczos, the claimed "exacitude" of the method is in fact not completely wrong.
The ED method involves no temperature statistics and no thermodynamical partition sums, hence it describes the system's physical properties at zero temperature. (However, there are recent extensions of ED working at finite temperature $T > 0$ [66]). Therefore, ED is a good complementary approach to other lattice simulation methods in many-particle physics, particularly the important finite-temperature QMC methods, which work efficiently at high temperatures but run into severe numerical problems if T tends to zero. Another advantage of ED is its ability to calculate dynamical response functions describing the system's reaction to external perturbations. Formally, these dynamical response functions are expressed by imaginary parts of Green functions [67] of the form

$$A(\mathbf{k},\omega) = \Im \frac{1}{\pi} \langle E_n | \hat{O}^\dagger(\mathbf{k}) \frac{1}{\omega - (H - E_n) - \mathrm{i}0_+} \hat{O}(\mathbf{k}) | E_n \rangle \,, \tag{2.1}$$

with $\mathbf{k}$ being the wave number and ω the angular frequency of the particle or quasiparticle associated with the perturbation. Theoreticians define $\hbar = 1$ for convenience, so that $\mathbf{k}$ becomes the momentum and ω the energy of the excitation. $\hat{O}(\mathbf{k})$ stands for a $\mathbf{k}$-dependent operator, and $| E_n \rangle$ is the n-th eigenstate of the system's Hamiltonian H with energy E_n. By inserting a $id = \left(\sum_j | E_j \rangle \langle E_j | = id \right)$ before and after the central fraction in Eq. (2.1) and using the identity

$$\frac{1}{x + \mathrm{i}0^+} = \mathcal{P}\left(\frac{1}{x}\right) - \mathrm{i}\pi\delta(x),$$

we obtain an alternative form of Eq. (2.1) which will prove more suitable for exploitation within the ED framework:

$$A(\mathbf{k},\omega) = \sum_j \left|\langle E_j|\hat{O}(\mathbf{k})|E_n\rangle\right|^2 \delta(\omega - (E_j - E_n)). \tag{2.2}$$

Dynamical response functions, which are also called *spectral functions* or *spectral densities*, are extraordinarily important for the theoretical description of quantum mechanical many-particle systems: they describe possible excitations and their dynamics and correspond to a vast number of experimetally accessible observables such as optical excitation energies, photoemission spectra or spin-spin correlation spectra. [67, 68].
If the operator $\hat{O}(\mathbf{k})$ within the Green function creates or annihilates one single particle, the corresponding Green function is called a *single-particle function.* A typical example is the photoemission of electrons from a solid (PE) or the inverse process, capturing a free electron accompanied by the emission of a photon (inverse photoemission, IPE) [68]:

$$A^{PE}(\mathbf{k},\omega) = \Im\frac{1}{\pi}\langle\psi|\, c^\dagger_{\mathbf{k},\sigma}\frac{1}{\omega+(H-E_\psi)-\mathrm{i}0_+}c_{\mathbf{k},\sigma}|\,\psi\rangle, \tag{2.3}$$

$$A^{IPE}(\mathbf{k},\omega) = \Im\frac{1}{\pi}\langle\psi|\, c_{\mathbf{k},\sigma}\frac{1}{\omega-(H-E_\psi)-\mathrm{i}0_+}c^\dagger_{\mathbf{k},\sigma}|\,\psi\rangle. \tag{2.4}$$

In *two-particle functions*, $\hat{O}(\mathbf{k})$ consists of two single-particle creators or annihilators. Important examples are the (dynamical) density-density and spin-spin correlation. The first quantity describes the dynamics of a momentum excitation in the studied many-particle system, the second reflects the system's response to a spin excitation connected with some momentum transfer – for example in a neutron scattering experiment: [68]

$$A^{DC}(\mathbf{k},\omega) = \Im\frac{1}{\pi}\langle\psi|n_{-\mathbf{k}}\frac{1}{\omega-(H-E_\psi)-\mathrm{i}0_+}n_{\mathbf{k}}|\,\psi\rangle\,, \tag{2.5}$$

$$A^{SC}(\mathbf{k},\omega) = \Im\frac{1}{\pi}\langle\psi|S^z_{-\mathbf{k}}\frac{1}{\omega-(H-E_\psi)-\mathrm{i}0_+}S^z_{\mathbf{k}}|\,\psi\rangle\,. \tag{2.6}$$

(Here it should be noted that $n^\dagger_{\mathbf{k}} = n_{-\mathbf{k}}$ because of $n_{\mathbf{k}} = \sum_{j=0}^{N_o-1} e^{-\mathrm{i}\mathbf{k}\mathbf{R}_j}n_j$, and analogously $S^{z\dagger}_{\mathbf{k}} = S^z_{-\mathbf{k}}$.)
Unfortunately, most numerical techniques in many-particle physics have rather severe problems in calculating these dynamical response functions. In QMC methods, for example, $A(\mathbf{k},\omega)$ has to be calculated from an inverted Laplace transformation; this only succeeds via a delicate Maximum Entropy restoration method [69, 70] with difficult error control and sometimes dissatisfying results. In summary, ED has its strengths where other methods run into problems: at low temperatures $T \to 0$ and when high-precision calculations of arbitrary spectral functions are needed. The restriction to very small system sizes, on the other hand, is a severe problem within ED. There are, however, two classes of system geometries in which these restrictions are less important: low dimensional systems like atomic chains or ladders, and single unit cells of a certain complexity, for example unit cells of cuprate superconductors or magnetic materials consisting of a couple of metal orbitals and a few non-metal ligand orbitals [14]. Furthermore, recent cluster expansion techniques have shown an interesting way out of the limited-size dilemma (see section 1.2.4): in this approach the physics of a small cluster is treated exactly, e.g. using ED, and then a large or even infinite number of identical clusters is coupled perturbatively to a macroscopically large system.

2.2 The Lanczos method

2.2.1 Algorithm

The Lanczos method [48, 71] is basically an algorithm which transforms a given Hermitean $(N\times N)$ matrix H into a tridiagonal matrix by performing N successive orthogonal transformations $H \to U^H H U$ with unitary matrices U. In other words, the initial basis $\mathcal{B}_0$ is transformed step by step into a new orthonormal basis $\mathcal{B} = \{|b_0\rangle, \ldots, |b_N\rangle\}$ in which the matrix H adopts a particularly simple form. In this procedure the first new basis vector $|b_0\rangle$ can be chosen arbitrarily, the following $|b_n\rangle$ are then calculated iteratively:

$$\begin{aligned} |b_1\rangle &= H|b_0\rangle - \frac{\langle b_0|H|b_0\rangle}{\langle b_0|b_0\rangle}|b_0\rangle \\ |b_{n+1}\rangle &= H|b_n\rangle - \frac{\langle b_n|H|b_n\rangle}{\langle b_n|b_n\rangle}|b_n\rangle - \frac{\langle b_n|b_n\rangle}{\langle b_{n-1}|b_{n-1}\rangle}|b_{n-1}\rangle, \quad n \geq 1. \end{aligned} \tag{2.7}$$

By means of complete induction it is easy to prove that the $|b_n\rangle$ constructed by Eq. (2.7) are indeed orthogonal by pairs. The coefficients appearing in the iteration equation shall be called α_n and β_n in the following:

$$\alpha_n := \frac{\langle b_n|H|b_n\rangle}{\langle b_n|b_n\rangle} \quad \text{and} \quad \beta_n^2 := \frac{\langle b_n|b_n\rangle}{\langle b_{n-1}|b_{n-1}\rangle}. \tag{2.8}$$

In the literature the Lanczos iteration equation is often presented in the form of Eq. (2.7) (see, for example [1]). However, this form is not very suitable for numerical application: if the absolute values of the eigenvalues of H are on average much larger (or much smaller) than 1, the norm of the $|b_n\rangle$ rapidly increases (decreases) in each iteration step, leading to numerical instabilities and a range overflow (underflow) after a certain number of steps. This problem can be avoided if at the end of each iteration step the new basis vector $|b_{n+1}\rangle$ is normalized:

$$|b_{n+1}\rangle' := \frac{|b_{n+1}\rangle}{\sqrt{\langle b_{n+1}|b_{n+1}\rangle}}. \tag{2.9}$$

From the definition of β_n we now obtain

$$\Big(\prod_{i=1}^{n}\beta_i\Big)|b_n\rangle' = \sqrt{\langle b_n|b_n\rangle}\,|b_n\rangle' = |b_n\rangle. \tag{2.10}$$

Inserting this into Eq. (2.7) and canceling out $\prod_{i=1}^{n}\beta_n$ on both sides of the equation gives us the Lanczos iteration equation in normalized vectors

$$\beta_{n+1}|b_{n+1}\rangle' = (H - \alpha_n)|b_n\rangle' - \frac{\beta_n^2}{\beta_n}|b_{n-1}\rangle', \tag{2.11}$$

which can be cast into an algorithmic form as follows:

$$\begin{aligned} |b_{n+1}\rangle' &\longleftarrow (H - \alpha_n)|b_n\rangle' - \beta_n|b_{n-1}\rangle' \\ \beta_{n+1} &\longleftarrow \sqrt{{}'\langle b_{n+1}|b_{n+1}\rangle'} \\ |b_{n+1}\rangle' &\longleftarrow |b_{n+1}\rangle' \,/\, \beta_{b+1}. \end{aligned}$$

From Eq. (2.11) and the orthonormality of the $|b_n\rangle'$ we obtain the matrix elements

$$\begin{aligned} H_{n,n} &= {}'\langle b_n|H|b_n\rangle' &&= \alpha_n, \\ H_{n,n-1} &= {}'\langle b_n|H|b_{n-1}\rangle' &&= \beta_n, \\ H_{n,m} &= {}'\langle b_n|H|b_m\rangle' &&= 0 \quad \text{für} \quad |n-m|>1. \end{aligned} \tag{2.12}$$

We see that the matrix is indeed transformed into diagonal form:

$$H = \begin{pmatrix} \alpha_0 & \beta_1 & & & \\ \beta_1 & \alpha_1 & \beta_2 & & \\ & \beta_2 & \alpha_2 & \beta_3 & \\ & & \beta_3 & \alpha_3 & \ddots \\ & & & \ddots & \ddots \end{pmatrix}. \tag{2.13}$$

Once H being tridiagonalized, the eigenvalue problem can easily be solved by standard routines from numerical libraries (e.g. the routine `stevx` from the `lapack` library[1]. The ED program codes described below use hand-coded routines implementing the method of Sturm's chains and bisection; this approach has the advantage to be particularly good-natured in situations in which some of the eigenvalues are almost degenerate, i.e. if there are 'clusters' of several closely neighbored eigenvalues. [72, 73]

2.2.2 Approximate solution of the eigenvalue problem

The Lanczos algorithm described above seems not suitable for the solution of very large eigenvalue problems: it needs a computational effort of $\mathcal{O}(N^3)$ operations and is therefore not better than other tridiagonalization methods, e.g. the Householder algorithm [72, 73]. However, the Lanczos method has one decisive advantage: under quite general suppositions (Kaniel-Paige theory, see [74–76]) one needs only $m \approx 100$ Lanczos transformation steps to obtain an $m \times m$ tridiagonal matrix $\langle b_i|H|b_j\rangle_{0\leq(i,j)<m}$ whose largest and lowest eigenvalues coincide extremely well with the largest and lowest eigenvalues of the full $N \times N$ matrix H – even for $N \approx 10^6$ or larger. The most important condition for this convergence of the largest and lowest eigenvalues E_i is that the initial vector $|b_0\rangle$ must not be orthogonal to one of the corresponding eigenvectors $|E_i\rangle$. In practice, this condition can be fulfilled (with a probability $p \approx 1$) by choosing the components of $|b_0\rangle$ randomly. The Lanczos algorithm is, therefore, not only an exact procedure of complexity $\mathcal{O}(N^3)$ for the solution of the eigenvalue problem but also an approximative method with complexity $\mathcal{O}(m\,N^2)$. Furthermore, the Hamiltonian matrices of typical many-particle problems in solid state physics are extremely sparsely populated – with typically $n = \mathcal{O}(N_o) < 100$ non-vanishing matrix elements per row. In this case, the computational effort even reduces to $\mathcal{O}(n\,m\,N)$ operations, which permits to attack eigenvalue problems for Hilbert space dimensions of up to $N \approx 10^9$ on modern supercomputers. For physical systems with four possible occupations per orbital this Hilbert space dimension corresponds to about 20 orbitals at half-filling, for systems with the no-double-occupancy constraint (like the

[1] further information about this freely available library can be obtained from the WWW site `http://www.netlib.org/lapack/lug/lapack_lug.html`.

t-J model, see section 1.1.2), 32 orbitals can be treated, and for Heisenberg-like models with 2 allowed states per orbital up to about 40 orbitals.
Besides computational complexity, the limited size of main memory can restrict the accessible system size in the Lanczos algorithm. Since one single complex vector $|b_n\rangle$ of length $N \approx 10^8$ requires about 2 GB of memory, it is evident that even on supercomputers it might not be possible to store all $m \approx 100...200$ vectors $|b_n\rangle$ simultaneously. Fortunately, there exists an easy way to circumvent this problem: noting that each Lancos iteration equation (2.7) connects three vectors $|b_{n+1}\rangle$, $|b_n\rangle$ and $|b_{n-1}\rangle$, we remark that one only needs to store three vectors if $|b_{n-2}\rangle$ is overwritten by $|b_{n+1}\rangle$ in each iteration step. Additionally, the initial vector $|b_0\rangle$ can be written to hard disk once the interation has been started. We then store the $m \approx 100...200$ Lanczos coefficients α_n and β_n and diagonalize the $(m \times m)$ tridiagonal matrix formed by the α_n and β_n. This provides us the m coefficients c_i of the expansion

$$|E\rangle = \sum_{i=0}^{m-1} c_i |b_i\rangle$$

of the desired energy eigenvector $|E\rangle$ with respect to the basis $\{|b_n\rangle\}$.
Now we restart the Lanczos procedure a second time with the same initial vector $|b_0\rangle$ (which has to be reloaded from disk) and we initialize $|E\rangle$ with $c_0|b_0\rangle$. Then we add a new term to $|E\rangle$ after each Lanczos step:

$$|E\rangle \rightarrow |E\rangle + c_i|b_i\rangle, \quad i = 1 \ldots m-1.$$

In this variant, the Lanczos algorithm requires only the memory to store 3 long vectors $|b_n\rangle$ – at the cost of doubling the total computation time. Using this approach, Hamiltonian matrices with dimensions up to 10^9 or even larger can be 'diagonalized' on modern supercomputers.

2.2.3 Dynamical response functions

The physical importance of response functions

$$A(\omega) = \frac{1}{\pi} \Im \langle E_n | \hat{O}^\dagger \frac{1}{\omega - (H - E_n) - \mathrm{i}0^+} \hat{O} | E_n \rangle \tag{2.14}$$

(with $\hat{O}$ being some quantum mechanical operator and $|E_n\rangle$ an energy eigenstate) has been discussed in section 2.1. Now we are going to show how the poles of $A(\omega)$ can be computed using the Lanczos algorithm. To this purpose we start the algorithm not with an arbitrary vector $|b_0\rangle$ but with the (normalized) vector

$$|b_0\rangle = \frac{\hat{O}|E_n\rangle}{\langle E_n|\hat{O}^\dagger\hat{O}|E_n\rangle^{1/2}}. \tag{2.15}$$

Starting from this $|b_0\rangle$ one obtains in $N-1$ Lanczos iterations the orthonormal Hilbert space basis $\{|b_0\rangle, \ldots, |b_{N-1}\rangle\}$. Next, we define $z := \omega + E_n - \mathrm{i}0^+$ and consider – following Fulde [77] – the identity

$$(z - H)\frac{1}{z - H} = id\,.$$

This identity can be inflated to

$$\sum_{n=0}^{N-1} \langle b_m | (z-H) | b_n \rangle \langle b_n | \frac{1}{z-H} | b_0 \rangle = \delta_{m,0}. \tag{2.16}$$

The matrix element on the left hand side can be calculated from the Lanczos iteration equation (2.7):

$$\langle b_m | (z-H) | b_n \rangle = (z-\alpha_n)\delta_{n,m} - \delta_{n+1,m} - \beta_n^2 \delta_{n-1,m},$$

so that (2.16) can be written as a tridiagonal matrix equation (with $x_n := \langle b_n | \frac{1}{z-H} | b_0 \rangle$):

$$\begin{pmatrix} z-\alpha_0 & -\beta_1^2 & & & \\ -1 & z-\alpha_1 & -\beta_2^2 & & \\ & -1 & z-\alpha_2 & \ddots & \\ & & \ddots & \ddots & -\beta_{N-1}^2 \\ & & & -1 & z-\alpha_{N-1} \end{pmatrix} \begin{pmatrix} x_0 \\ x_1 \\ x_2 \\ \vdots \\ x_{N-1} \end{pmatrix} = \begin{pmatrix} 1 \\ 0 \\ 0 \\ \vdots \\ 0 \end{pmatrix}. \tag{2.17}$$

The variable $x_0 = \langle b_0 | \frac{1}{z-H} | b_0 \rangle$ is the desired quantity in this system. In order to extract it, we eliminate the other x_n step by step from bottom to top:

$$-x_{N-2} + (z-\alpha_{N-1})x_{N-1} = 0 \;\Rightarrow\; x_{N-1} = \frac{x_{N-2}}{z-\alpha_{N-1}}$$

$$-x_{N-3} + (z-\alpha_{N-2})x_{N-2} - \beta_{N-1}^2 x_{N-1} = 0 \;\Rightarrow\; x_{N-2} = \cfrac{x_{N-3}}{z - a_{N-2} - \cfrac{\beta_{N-1}^2}{z-\alpha_{N-1}}}$$

$$\vdots \qquad\qquad \vdots$$

Finally, we obtain x_0 in the form of a chain fraction consisting of the Lanczos coefficients α_n and β_n. For $A(\omega)$ we get the equation

$$A(\omega) = -\frac{1}{\pi}\Im \left(\cfrac{\langle E_n | \hat{O}^\dagger \hat{O} | E_n \rangle}{z-\alpha_0 - \cfrac{\beta_1^2}{z-\alpha_1 - \cfrac{\beta_2^2}{z-\alpha_2 - \cdots}}} \right) \tag{2.18}$$

with which the positions of the poles of $A(\omega)$ can be calculated from $\{\alpha_n\}$ and $\{\beta_n\}$. The poles' intensities are the weights $|\langle E_j | \hat{O} | E_n \rangle|^2$ in the alternative form (2.2) of $A(\omega)$; they are proportional to the overlap of the energy eigenstates $| E_j \rangle$ with the starting state $| b_0 \rangle$:

$$|\langle E_j | \hat{O} | E_n \rangle|^2 = |\langle E_j | b_0 \rangle|^2 \langle E_n | \hat{O}^\dagger \hat{O} | E_n \rangle. \tag{2.19}$$

In summary, we have shown that all information contained in the dynamical response function $A(\omega)$ can be accessed using the Lanczos method.

2.3 Efficient implementation

2.3.1 Coding the basis states

In the formalism of Second Quantization [10], quantum mechanical many-particle states of n particles (fermions or bosons) are constructed by applying n particle creating operators $c^\dagger_{i,\sigma}$ to the vacuum state $|\, vac\,\rangle$:

$$|\,\psi\,\rangle = c^\dagger_{i_n,\sigma_n} \cdots c^\dagger_{i_1,\sigma_1} \,|\, vac\,\rangle \tag{2.20}$$

(i is the orbital index, $\sigma = \uparrow, \downarrow$ the spin index).
Since fermion creating operators *anti*commute [10],

$$\left[c^\dagger_{i,\sigma},\, c^\dagger_{j,\tau}\right]_+ = 0 = \left[c_{i,\sigma},\, c_{j,\tau}\right]_+ \quad \text{and} \quad \left[c_{i,\sigma},\, c^\dagger_{j,\tau}\right]_+ = \delta_{i,j}\delta_{\sigma,\tau}, \tag{2.21}$$

one needs two informations to unambiguously characterize the quantum mechanical state of a fermionic many-particle system such as n interacting electrons:

1. How many particles are there in the system and what are their orbital and spin quantum numbers, i.e. which fermion creators have to be applied to the vacuum state?

2. In which order are the fermion creators applied?

The best way to code fermionic many-particle states in a computer is to use a bit field in which each bit stands for one allowed combination of orbital and spin indices of the system. Bits equal to 1 indicate that the system contains a fermion with the corresponding orbital and spin quantum numbers, bits equal to 0 indicate that no particle with these orbital and spin values exists in the system. The state $|{\uparrow} - {\downarrow} {\uparrow\!\downarrow}\rangle$ of an electron system with four orbitals, for example, could then be represented as

$$|{\uparrow} - {\downarrow} {\uparrow\!\downarrow}\rangle \longrightarrow |\,1001;0011\,\rangle \longrightarrow |\,9;3\rangle,$$

where the first four bits symbolize the presence of spin-up electrons, the last four bits mark the presence of spin-down electrons. The problem of ordering the creation operators can be solved by introducing a standard order: when writing a bit representation of a state such as $|\,1001;0011\,\rangle$, we implicitly assume that this state has been constructed from the vacuum state by applying a series of creation operators in which the operators have the same relative positions as the 1-bits created by them in the bit field. In our example, this means

$$|{\uparrow} - {\downarrow} {\uparrow\!\downarrow}\rangle \equiv |\,1001;0011\,\rangle := c^\dagger_{1,\uparrow}\, c^\dagger_{4,\uparrow}\, c^\dagger_{3,\downarrow}\, c^\dagger_{4,\downarrow}\,|\, vac\,\rangle\,.$$

When applying an arbitrary operator consisting of creation and/or annihilation operators $c^\dagger_{i,\sigma}$ and/or $c_{j,\tau}$ to a many-particle state like $|{\uparrow} - {\downarrow} {\uparrow\!\downarrow}\rangle$, we are faced with the problem of how to merge the operator's creators and annihilators into the sequence of existing creators defining the given state. In order to maintain unambiguity we have to reconstruct the standard order, i.e. the new operators – initially appended at the left side of the operator sequence defining the state – have to be moved to the right till they reach a position which is consistent with standard order. This is done by repeatedly swapping the positions of two adjacent fermion operators, each of these commutations introducing an additional prefactor of -1 due to the fermion commutation relations (2.21). Finally, this results in a so called *fermi sign* prefactor of +1 resp. -1 if the total number of operator commutations is even resp. odd. In our example state, the hopping of a spin-down electron from orbital 4 to orbital 3 – described by the operator $c^\dagger_{3,\downarrow}\, c_{4,\downarrow}$ – has a fermi sign of -1 because five operator commutations are needed to insert $c^\dagger_{3,\downarrow}$ and $c_{4,\downarrow}$ correctly:

$$c^\dagger_{2,\downarrow}\, c_{4,\downarrow} | \,\mathrlap{\uparrow}{-} - \mathrlap{\downarrow}{-} \,\uparrow\!\downarrow\rangle = c^\dagger_{2,\downarrow}\, c_{4,\downarrow}\, c^\dagger_{1,\uparrow}\, c^\dagger_{4,\uparrow}\, c^\dagger_{3,\downarrow}\, c^\dagger_{4,\downarrow}\, |\, vac\,\rangle = (-1)^5\, c^\dagger_{1,\uparrow}\, c^\dagger_{4,\uparrow}\, c^\dagger_{2,\downarrow}\, c^\dagger_{3,\downarrow}\, |\, vac\,\rangle\,.$$

A more detailed discussion on efficient administration of Hilbert basis states and implementation tricks for the calculation of fermi signs and operator application can be found in [65].

2.3.2 Reducing Hilbert space dimension using symmetries

One of the most limiting problems of the ED technique is the exponential growth of Hilbert space dimension with the number of orbitals and particles in the system. Fortunately, this dimension can be reduced considerably by exploiting the systems inherent symmetries and invariances in a clever way [79, 80]. A physical observable $\hat{G}$ is an *invariant* of the system if it commutes with the Hamiltonian [10], i.e. if

$$[H, \hat{G}]_- = 0. \tag{2.22}$$

If we now assume that $|\, g\,\rangle$ is an eigenstate of $\hat{G}$ with eigenvalue g and if equation (2.22) holds, then $H|\, g\,\rangle$ is also an eigenstate of $\hat{G}$ with the same eigenvalue, which can be seen quite easily from the following transformation:

$$\hat{G}(H|\, g\,\rangle) = (\hat{G}H)|\, g\,\rangle = (H\hat{G})|\, g\,\rangle = H(g|\, g\,\rangle) = g(H|\, g\,\rangle).$$

As a consequence, the Hamiltonian matrix adopts a block diagonal form if one uses a Hilbert basis of $\hat{G}$-eigenstates; each diagonal block contains the subspace of all $\hat{G}$-eigenstates with one fixed eigenvalue g. These blocks can be diagonalized independently, which has two important advantages: on the one hand it leads to much smaller matrix dimensions, and on the other hand it allows to specify the physical properties of the system's starting state more precisely.
On regularly ordered orbital geometries (lattices) with *periodic boundary conditions*, typical Hamiltonians such as the Hubbard, t-J or Heisenberg model (see section 1.1) have some or all of the following symmetries and invariances:

- particle number conservation (N).
- spin rotation symmetry $\longrightarrow$ conservation of total spin and z-spin ($\mathbf{S}^2$ and S_z).
- translational symmetry $\longrightarrow$ momentum conservation ($\mathbf{p} = \hbar\mathbf{k}$).
- point symmetry group of the finite lattice $\longrightarrow$ invariance under certain inversions and rotations (the corresponding conserved quantities are parity P and the quantum number of the discrete lattice rotation, which can be considered as a discrete spatial angular momentum).

- In the $(S_z = 0)$ subspace: reflection symmetry along the z-axis $\longrightarrow$ spin inversion invariance.
- At half filling: particle-hole symmetry.

If a certain symmetry is to be used to split up the H matrix into smaller blocks, one first has to find an eigenbasis of the quantum mechanical operator connected with this symmetry. This is very easy for some operators such as total particle number $\hat{N}$ or the total z-spin $\hat{S}_z$, but extremely difficult for others like the total spin $\mathbf{S}^2$. Finding an eigenbasis of $\mathbf{S}^2$ is for most models almost as difficult as the original task, diagonalizing the Hamiltonian itself, so that in most practical applications this symmetry cannot be used to reduce the Hilbert space dimension. The practically most important symmetries and invariances are the conservation of N and S_z as well as momentum conservation due to translational invariance in systems with periodic boundary conditions. Exploiting these symmetries will be the issue of the following subsections. A more detailed discussion together with pseudocode algorithms can be found in [65, 78].

Particle number and z-spin eigenbasis

The most commonly used Hilbert basis for representing many-particle systems is a product basis of single-orbital eigenbases of $\hat{N}$ and $\hat{S}_z$. We have used this approach when writing down states like $|\uparrow - \downarrow \uparrow\!\downarrow\rangle$ in subsection 2.3.1. Hence, the full many-particle basis is already an eigenbasis of $\hat{N}$ and $\hat{S}_z$, and the desired new basis with fixed values of N and S_z can be created by just throwing away all initial basis states with wrong N and/or S_z numbers. For systems with 4 allowed states per orbital (such as the Hubbard model), the dimension of Hilbert space thus reduces from 4^{N_o} to

$$dim_{N,S_z} = \binom{N_o}{n_\uparrow}\binom{N_o}{n_\downarrow} \quad \text{with} \quad n_\uparrow = \frac{N}{2} + S_z,\; n_\downarrow = \frac{N}{2} - S_z \tag{2.23}$$

(with N_o being the total number of orbitals in the system).

Momentum eigenbasis

Constructing a Hilbert basis of momentum eigenstates is not as straight forward as constructing the $(\hat{N}, \hat{S}_z)$ eigenbasis in the previous subsection. The procedure corresponds to the construction of *Bloch* states (plain waves) $|\,\mathbf{k}\,\rangle$ starting from spatially localized *Wannier* states $|\,\mathbf{r}\,\rangle$ [5, 10]:

$$|\,\mathbf{k}\,\rangle = A_{\mathbf{k}} \sum_{j=0}^{N_o-1} \hat{\mathcal{T}}_{\mathbf{k}(\mathbf{R}_j)} |\,\mathbf{r}\,\rangle = A_{\mathbf{k}} \sum_{j=0}^{N_o-1} e^{\mathrm{i}\mathbf{k}\mathbf{R}_j} |\,\mathbf{r} + \mathbf{R}_j\,\rangle. \tag{2.24}$$

Here, $\hat{\mathcal{T}}_{\mathbf{k}}$ is the translation operator and the $\mathbf{R}_j$ symbolize all allowed translations mapping the chosen finite lattice onto itself (thereby taking into account the periodic boundary conditions). The $|\,\mathbf{r}\,\rangle$ can be identified with the states of the $(\hat{N}, \hat{S}_z)$ eigenbasis. The set of allowed $\mathbf{k}$ (eigen-)values is finite and discrete just as the set of valid lattice points $\mathbf{r}_j$ [5]:

$$\mathbf{k} = \Big(\frac{2\pi n_x}{N_x}, \frac{2\pi n_y}{N_y}, \frac{2\pi n_z}{N_z}\Big) \quad \text{with} \quad 0 \le n_{x/y/z} < N_{x/y/z}, \tag{2.25}$$

where $N_{x/y/z}$ is the number of orbitals in $x/y/z$-direction (and therefore $N_x N_y N_z = N_o$). When calculating the translated states $\hat{T}_{\mathbf{k}(\mathbf{R}_j)}|\,\mathbf{r}\,\rangle$, we have to take into account eventual fermi signs which are introduced by the reordering of the state's particle creating operators due to translation and periodic boundary conditions. In a one dimensional two-site system, for example, we get

$$\hat{T}_{k=0(1)}|\uparrow\uparrow\rangle = -|\uparrow\uparrow\rangle\,,$$

because the translation effectively swaps the two spin-up electrons in the system.
Coding the $\mathbf{k}$-basis states in a computer is most conveniently done by storing *one* $\mathbf{r}$-state of the expansion (2.24) as a representative for each translationally invariant state. One convenient choice is to take always the $\mathbf{r}$-state with the smallest bit field value in the expansion (2.24).
Additionally, the Bloch states' norm prefactors $A_{\mathbf{k}}$ have to be stored. At first glance it seems that the prefactors are all identical and equal to $\frac{1}{\sqrt{N_o}}$, but this is not the case for $\mathbf{k}$-states with particular symmetries, resulting in the property that the representative is identically reproduced after less than N_o translations. If the N_o different translations of the representative produce N_{diff} different $\mathbf{r}$-states – each of them $N_{id} = N_o/N_{diff}$ times – then the norm prefactor is

$$A_{\mathbf{k}} = \frac{1}{\sqrt{N_o}\sqrt{N_{id}}}. \tag{2.26}$$

The $\mathbf{k}=\mathbf{0}$-state constructed from the state $|\uparrow\downarrow\uparrow\downarrow\rangle$ of a four-site chain, for example, has the prefactor

$$A_0 \sum_{j=0}^{3} \hat{T}_{0(\mathbf{R}_j)}|\uparrow\downarrow\uparrow\downarrow\rangle = A_0\big(2\,|\uparrow\downarrow\uparrow\downarrow\rangle + 2\,|\downarrow\uparrow\downarrow\uparrow\rangle\big) \quad \Rightarrow \quad A_0 = \frac{1}{2\sqrt{2}}.$$

However, only a very small minority of all $\mathbf{r}$-states possesses such a particular symmetry; for most $\mathbf{r}$-states we have $N_{diff} = N_o$. Therefore, the dimension of the $(\hat{N}, \hat{S}_z, \hat{\mathbf{k}})$ eigenbasis is roughly N_o times smaller than the dimension of the $(\hat{N}, \hat{S}_z)$ basis:

$$dim_{N,S^z,\mathbf{k}} \gtrsim \frac{1}{N_o}\binom{N_o}{n_\uparrow}\binom{N_o}{n_\downarrow}. \tag{2.27}$$

Calculating $|\,y\,\rangle = H|\,x\,\rangle$in the momentum eigenbasis

The momentum eigenbasis has been introduced in order to split up the Hamiltonian into smaller and numerically more convenient blocks. However, it seems that the calculation of a single matrix element

$$\langle\,\mathbf{k}_2\,|H|\,\mathbf{k}_1\,\rangle \tag{2.28}$$

needs N_o^2 times more operations than the corresponding calculation in the $(\hat{N}, \hat{S}_z)$ eigenbasis because the expansion (2.24) has to be performed both on the bra and the ket side of the matrix element. Fortunately, it can be shown that instead of calculating (2.28) one can as well compute

$$\langle\,r_2\,|H|\,r_1\,\rangle\,, \tag{2.29}$$

with $|\mathbf{r}_1\rangle$ and $|\mathbf{r}_2\rangle$ being $\mathbf{r}$-space representatives for $|\mathbf{k}_1\rangle$ and $|\mathbf{k}_2\rangle$. This can be seen quite easily by inserting the expansion

$$|\mathbf{k}_i\rangle = \frac{1}{\sqrt{N_o N_{id}}} \sum_{\nu=0}^{N_o-1} \mathcal{T}_{\mathbf{k}(\mathbf{R}_\nu)} |r_i\rangle$$

into equation (2.28):

$$\langle \mathbf{k}_2 | H | \mathbf{k}_1 \rangle = \frac{1}{N_o \sqrt{N_{id}(r_1) N_{id}(r_2)}} \sum_{\mu=0}^{N_o-1} \sum_{\nu=0}^{N_o-1} \langle r_2 | \mathcal{T}^\dagger_{\mathbf{k}(\mathbf{R}_\mu)} H \, \mathcal{T}_{\mathbf{k}(\mathbf{R}_\nu)} | r_1 \rangle. \tag{2.30}$$

Due to $\left[\hat{\mathbf{k}}, H\right]_- = 0$, the term $\mathcal{T}_{\mathbf{k}(\mathbf{R})} = e^{-i\hat{\mathbf{k}}\mathbf{R}}$ commutes with H. Furthermore, the translation operator has the following properties:

$$\mathcal{T}^\dagger_{\mathbf{k}(\mathbf{R})} = \mathcal{T}_{\mathbf{k}(-\mathbf{R})} \quad \text{and} \quad \mathcal{T}_{\mathbf{k}(\mathbf{R}_1)} \mathcal{T}_{\mathbf{k}(\mathbf{R}_2)} = \mathcal{T}_{\mathbf{k}(\mathbf{R}_1+\mathbf{R}_2)}.$$

Therefore, we get

$$\mathcal{T}^\dagger_{\mathbf{k}(\mathbf{R}_\nu)} H \, \mathcal{T}_{\mathbf{k}(\mathbf{R}_\mu)} = \mathcal{T}_{\mathbf{k}(-\mathbf{R}_\nu)} \mathcal{T}_{\mathbf{k}(\mathbf{R}_\mu)} H = \mathcal{T}_{\mathbf{k}(\mathbf{R}_\mu - \mathbf{R}_\nu)} H.$$

Now let $\tilde{\nu} := \nu - \mu$ be a new index variable. Using $\tilde{\nu}$ we can reformulate (2.30) as follows:

$$\begin{aligned}
\langle \mathbf{k}_2 | H | \mathbf{k}_1 \rangle &= \frac{1}{\sqrt{N_{id}(r_1) N_{id}(r_2)}} \left(\sum_{\mu=0}^{N_o-1} \frac{1}{N_o} \right) \sum_{\tilde{\nu}=0}^{N_o-1} \langle r_2 | \mathcal{T}_{\mathbf{k}(\mathbf{R}_{\tilde{\nu}})} H | r_1 \rangle \\
&= \frac{1}{\sqrt{N_{id}(r_1) N_{id}(r_2)}} N_{id}(r_2) \sum_{\tilde{\nu}=0}^{N_{diff}(r_2)-1} \langle r_2 | \mathcal{T}_{\mathbf{k}(\mathbf{R}_{\tilde{\nu}})} H | r_1 \rangle \\
&= \sqrt{\frac{N_{id}(r_2)}{N_{id}(r_1)}} \sum_{\tilde{\nu}=0}^{N_{diff}(r_2)-1} \mathrm{e}^{i\mathbf{k}\mathbf{R}_{\tilde{\nu}}} s^F_{-\tilde{\nu}} \langle T_{-\tilde{\nu}} r_2 | H | r_1 \rangle.
\end{aligned} \tag{2.31}$$

Here, $|T_{-\tilde{\nu}} r_2\rangle$ is the state $|r_2\rangle$ shifted by $-\mathbf{R}_{\tilde{\nu}}$, and $s^F_{-\tilde{\nu}}$ is the fermi sign originating from this translation.

In almost any reasonable microscopic model, the Hamiltonian H decomposes into a sum of different elementary interaction terms $H^{(\alpha)}$, such as hopping terms $t\, c^\dagger_{j,\sigma}\, c^\dagger_{i,\sigma}$ or Coulomb interactions $U n_j n_i$. Each term $H^{(\alpha)}$, when applied to one single $(\hat{N}, \hat{S}_z)$ basis state $|r_1\rangle$, returns the multiple of *one* new $(\hat{N}, \hat{S}_z)$ basis state $|\tilde{r}_1\rangle$. This is called 'non-branching property' of the Hamiltonian's elementary interactions:

$$H^{(\alpha)} |\mathbf{r}_1\rangle = s^F_\alpha h^{(\alpha)} |\tilde{\mathbf{r}}_1\rangle.$$

s^F_α is the fermi sign resulting from the standard ordered insertion of the fermionic operators contained in $H^{(\alpha)}$ into the sequence of operators defining the state $|\mathbf{r}_1\rangle$.

Thus, if one writes (2.31) for one single term $H^{(\alpha)}$ and not for the entire Hamiltonian, the sum over $\tilde{\nu}$ can have at most *one* non-vanishing term; the latter is found if and only if $|\tilde{r}_1\rangle = |T_{-\tilde{\nu}} \mathbf{r}_2\rangle$ holds. In this case one obtains

$$\langle \mathbf{k}_2 | H^{(\alpha)} | \mathbf{k}_1 \rangle = \sqrt{\frac{N_{id}(r_2)}{N_{id}(r_1)}} \mathrm{e}^{i\mathbf{k}\mathbf{R}_{\tilde{\nu}}} s^F_{-\tilde{\nu}} s^F_\alpha h^{(\alpha)}. \tag{2.32}$$

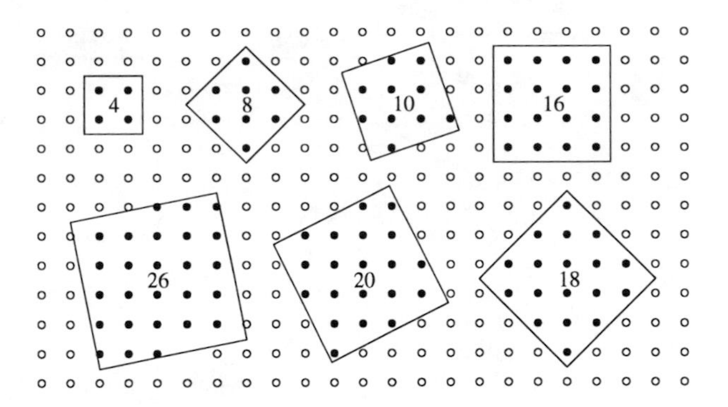

Figure 2.1: The smallest nontrivial square clusters with an even number of sites.

This proves the assertion made at the beginning of this subsection: matrix elements in the momentum eigenbasis can be computed by simply referring to the corresponding representative **r**-states. Further implementation tricks for working with the momentum eigenbasis can be found in [65].

2.3.3 Periodic lattices

Numerical studies of 'standard' microscopic models (such as the Hubbard, t-J or Heisenberg model) on perfectly ordered orbital lattices have been one of the main applications of ED methods in solid state physics. Due to the limited accessible orbital number within ED, one is in most cases restricted to studying low dimensional lattice geometries such as atomic chains, ladders or planes containing not more than a few dozens of orbitals (called 'sites'). The almost infinite number of atoms and orbitals in real world crystals is often simulated in ED studies by introducing periodic boundary conditions (PBC), i.e. by connecting the left boundary of the finite lattice to its right boundary and/or its lower edge to its upper one. A chain with PBC can be visualized as a closed ring and a plane with PBC in both directions can be regarded as a torus. In this subsection, we focus on two- imensional systems, in particular square lattices with PBC; these lattices possess a rich point symmetry group and are thus good demonstration objects for studying the methods of Hilbert space reduction by using symmetries. Figure 2.1 shows the seven smallest nontrivial square clusters containing an even number of sites. (Clusters with an odd number of sites can be unsuitable for lattice simulations since they lead to frustration if, for example, the system tends to adopt an antiferromagnetic or checkerboard order.) As can be seen from the figure, the spanning vectors of the square, $\mathbf{a} = (x, y)$ and $\mathbf{b} = (-y, x)$, need not be parallel to the lattice axes. The only criterion for x and y is

$$x, y \in \mathbb{N}_0 \quad \text{and} \quad x + y \quad \text{even.}$$

Table 2.1 lists some practically important square clusters on a 2D square lattice with their symmetry operations. The product of the number of translations, rotations and inversion

orbitals	spanning vector	translations	rotations	inversions
4	(2,0)	4	-	2
8	(2,2)	8	4	2
10	(3,1)	10	4	-
16	(4,0)	16	4	2
18	(3,3)	18	4	2
20	(4,2)	20	4	-
26	(5,1)	26	4	-
32	(4,4)	32	4	2
34	(5,3)	34	4	-
36	(6,0)	36	4	2
40	(6,2)	40	4	-
50	(5,5)	50	4	2
50	(7,1)	50	4	-
52	(6,4)	52	4	-
58	(7,3)	58	4	-
64	(8,0)	64	4	2

Table 2.1: Properties and symmetry operations of the smallest square lattices with periodic boundary conditions.

is roughly the factor about which the Hilbert space dimension can be reduced if all these symmetries are exploited.
One prerequisite for constructing eigenstates of lattice translation, rotation and inversion in a computer program is to have an algorithm which automatically numbers the sites of any square cluster in a consistent way. Figure 2.2 shows one possible solution to this problem: start at point $(0, 0)$ and move in the direction of spanning vector **a** until a new cluster point is reached, thereby taking into account the PBC. If this point has not been visited before, it is assigned the next free number and the move in **a**-direction continues. If, however, the point has been visited and labeled before, then an additional step in y-direction is performed. This algorithm guarantees that all points are labeled exactly once before returning to the starting point (0,0) after N_o steps.
In paragraph 2.3.2 we described the construction of translationally invariant momentum eigenstates $|\,\mathbf{k}\,\rangle$ from localized states $|\,\mathbf{r}\,\rangle$ by equation(2.24). The equivalent equation for rotation operations on the square cluster reads

$$|\,q\,\rangle = A_q \sum_{j=0}^{3} \hat{\mathcal{R}}(j\pi|\,0\,\rangle = A_q \sum_{j=0}^{3} \mathrm{e}^{\mathrm{i}\pi qj}|\,j\pi\,\rangle, \tag{2.33}$$

where $q = 0, \frac{1}{2}, 1$ or $\frac{3}{2}$ is the quantum number of rotation and $\hat{\mathcal{R}}(\phi)$ the rotation operator around the z-axis with angle ϕ. Analogously, an inversion eigenstate is constructed via

$$|\,p\,\rangle = A_p\,(|\,0\,\rangle + \hat{o}|\,0\,\rangle) = A_p\,(|\,0\,\rangle + (-1)^p|\,\mathrm{inv}\,\rangle)\,, \tag{2.34}$$

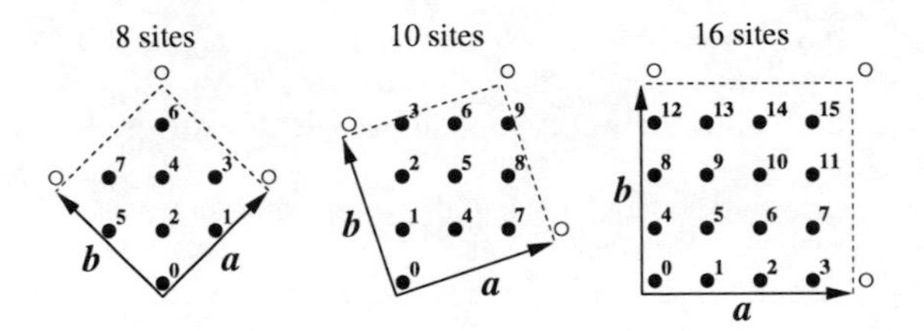

Figure 2.2: Some tilted square clusters with their spanning vectors **a**, **b**. *The sites are numbered by starting at (0,0) and by then repeatedly going in* **a**-*direction till a new site is reached. If one thus returns to a site visited before, one additionally performs a step upwards ($dy = +1$).*

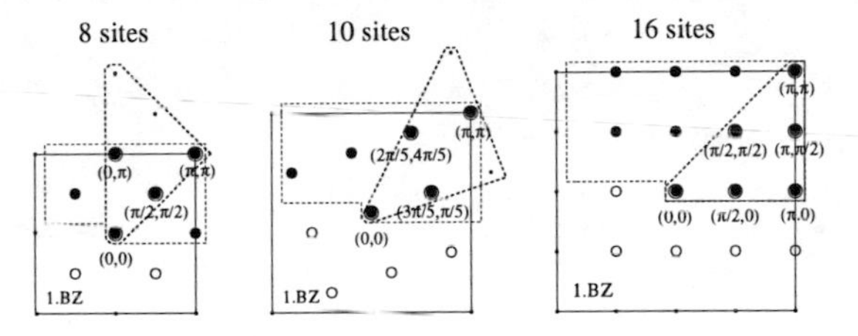

Figure 2.3: Allowed k-points in the first Brillouin zone (1.BZ) for the clusters shown in Fig. 2.1. There are N_{sites} k-points (open and filled circles). If one counts **k** *and* $-$**k** *only once one is left with the filled circles (upper half of 1.BZ). The topologically different k-points (double circles) are a subset of this set; it can be found by forming the intersection of the lines $0 + \lambda \cdot \mathbf{a}$ and $0 + \lambda \cdot (\mathbf{a} + \mathbf{b})$ with 1.BZ*

with $p=0$ or 1 being the parity quantum number and $\hat{o}$ the spatial inversion operator. The allowed range for the quantum numbers of rotation and inversion, $q \in \{0, \frac{1}{2}, 1, \frac{3}{2}\}$ and $p \in \{0, 1\}$, is easy to determine and identical for all square clusters. The allowed quantum numbers of translation, the momenta (or wave numbers) **k**, however, vary from cluster to cluster and are not trivial to find, in particular for the tilted clusters. It can be shown that if the spanning vectors of the square cluster are $\mathbf{a}_1 = (x, y)$ and $\mathbf{a}_2 = (-y, x)$, then the allowed **k**-values are multiples of two reciprocal lattice vectors $\mathbf{K}_1$ and $\mathbf{K}_2$,

$$\mathbf{k}_{m,n} = m\mathbf{K}_1 + n\mathbf{K}_2 \,, \quad m, n \in \mathbb{N}_0 \tag{2.35}$$

with

$$\mathbf{K}_1 = \begin{pmatrix} x \\ y \end{pmatrix} \frac{2\pi}{x^2 + y^2} \quad \text{and} \quad \mathbf{K}_2 = \begin{pmatrix} y \\ -x \end{pmatrix} \frac{2\pi}{x^2 + y^2}. \tag{2.36}$$

This follows from the reciprocal lattice condition $\mathbf{a}_i \cdot \mathbf{K}_j = 2\pi\delta_{i,j}$ [5]. Figure 2.3 shows the allowed **k**-points in the first Brillouin zone (1BZ) for the square clusters with 8, 10 and 16 sites.

The total number of **k**-points in the 1BZ is always equal to N_o, the total number of orbitals. However, many of these **k**-points are topologically equivalent in the sense that

they can be mapped onto each other by exchanging k_x and $-k_x$, k_y and $-k_y$, or k_x and k_y. The remaining subset of topologically different **k**-points can constructed by retaining only those **k**-points in the 1BZ which are situated in between the directions $\mathbf{K}_1$ and $\mathbf{K}_1 + \mathbf{K}_2$ (see figure 2.3).
We end this paragraph with a plot showing all symmetry operations of the 8-site square cluster together with a possible concatenation of translations, rotations and inversions creating all the configurations that can be reached starting from one "representative" (figure 2.4).

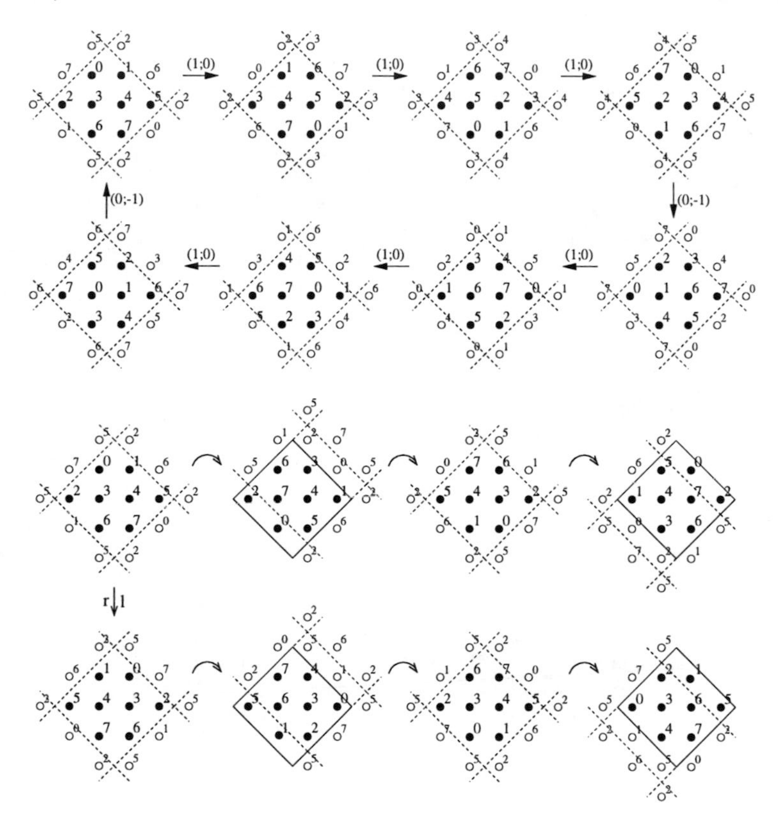

Figure 2.4: All translations (above), rotations and inversions (below) mapping the 8-site square cluster with periodic boundary conditions (PBC) onto itself.

2.3.4 Matrix-free methods for specific electron models

In practical ED applications the Hamiltonian matrices often have dimensions of up to 10^9. In these cases it might still be possible to store the three state vectors $|\, b_i \,\rangle$ needed for the Lanczos algorithm (see section 2.2), but it is definitely impossible to store the H matrix itself, even on the largest supercomputers. For this reason so-called matrix-free techniques are very important in ED implementations. In this approach, the elements of H are not stored but recomputed whenever they are needed to perform the matrix-vector multiplication

$$|\, b_{i+1}\,\rangle = H|\, b_i\,\rangle \tag{2.37}$$

which forms the central and most time consuming step within the Lanczos algorithm. This is done as follows: first the Hamiltonian is split up into elementary interaction terms h having the non-branching property (see section 2.3.2). Then the list of 'active' states has to be determined for each elementary term h, i.e. all those basis states on which h can be applied with non-vanishing result. For each of these active states the new state after application of h is calculated, as well as the Fermi sign resulting from this operation. The non-branching property of h guarantees that this new state is again – up to an additional prefactor – a basis state. Now a crucial step has to be performed: the position of the new state in the list of all Hilbert basis states has to be found in order to get the row index of the resulting matrix element of H.
One possible implementation for this search operation is a binary search in the (ordered) list of all basis states. This is not very convenient since it needs about $\log_2(\dim_H) \approx 30$ search steps and produces a lot of cache misses as different parts of a huge array of length $\dim_H$ have to be loaded into the cache memory. Another approach is to establish an inverted list: if the basis states are coded as bit fields of n bits, all states can be indexed using an inverted list of length 2^n. However, since $n \approx 40$ is a realistic value for large ED applications (e.g. the Hubbard model on 20 sites), it is clear that such an index array would be by far too large for efficient handling even on the largest available supercomputers.
Much better than the two ideas described above is a so-called *two-table* lookup scheme [81]: instead of creating one inverted list of length 2^n two partial indices are used and two inverted lists of roughly the length $2^{n/2}$ – which is typically not more than 10^4 to 10^5 – are stored. The full index is then computed as a combination of two partial indices which can be directly read from the two inverted lists.
In the following paragraphs some strategies will be presented on how to implement two-table lookup schemes together with Hilbert bases exploiting symmetries. We will see that there are three important classes of microscopic models of correlated electron systems; for each class separate implementation tricks are nessessary:

- *Hubbard-like models*, i.e. models with four allowed states per orbital ($|-\rangle$, $|\uparrow\rangle$, $|\downarrow\rangle$, and $|\uparrow\downarrow\rangle$).
- *t-J-like models*, i.e. models with three allowed states per orbital ($|-\rangle$, $|\uparrow\rangle$ and $|\downarrow\rangle$).
- *Heisenberg-like models*, i.e. models with two allowed states per orbital ($|\uparrow\rangle$ and $|\downarrow\rangle$).

Hubbard-like models

The Hubbard model [11] has been described in section 1.1.1. As 'Hubbard-like' models we want to summarize all microscopic many-electron models with the following two properties:

- The model's Hamiltonian conserves the total electron number and the electrons' total z-spin.
- The model acts on a set of orbitals with four allowed states per orbital: $|-\rangle$, $|\uparrow\rangle$, $|\downarrow\rangle$, and $|\updownarrow\rangle$.

In these models, the two-table lookup is most conveniently implemented by treating the spin-up and the spin-down electrons completely independently. In this section, we will describe how the two-table lookup can be implemented together with the $(N, S_z, \mathbf{k})$ eigenbasis. The generalization to eigenbases exploiting further symmetries, e.g. the point group symmetries of the cluster, is straight forward and should not impose any severe problems. At first let us introduce some notations: the (N, S_z) eigenbasis shall be called $\mathcal{B}_{N,S_z}$, the (N, S_z) eigenbasis $\mathcal{B}_{N,S_z,\mathbf{k}}$, and the corresponding partial bases for the subspaces containing only spin-up (spin-down) electrons $\mathcal{B}^{\uparrow}_{N,S_z}$ and $\mathcal{B}^{\uparrow}_{N,S_z,\mathbf{k}}$ ($\mathcal{B}^{\downarrow}_{N,S_z}$ and $\mathcal{B}^{\downarrow}_{N,S_z,\mathbf{k}}$). We note that the (N, S_z) eigenbasis can be written as a direct product of its two partial bases,

$$\mathcal{B}_{N,S_z} \equiv \mathcal{B}^{\uparrow}_{N,S_z} \otimes \mathcal{B}^{\downarrow}_{N,S_z}.$$

Our standard example in this section will be the 4-site chain with PBC carrying 2 spin-up and 2 spin-down electrons. For this system, $\mathcal{B}^{\uparrow}_{N,S_z}$ consists of six states,

$$\mathcal{B}^{\uparrow}_{N,S_z} = \{ \ |{-}{-}{\uparrow}{\uparrow}\rangle, |{-}{\uparrow}{-}{\uparrow}\rangle, |{-}{\uparrow}{\uparrow}{-}\rangle, |{\uparrow}{-}{-}{\uparrow}\rangle, |{\uparrow}{-}{\uparrow}{-}\rangle, |{\uparrow}{\uparrow}{-}{-}\rangle \},$$

and the same holds for $\mathcal{B}^{\downarrow}_{N,S_z}$ (with $\uparrow$ replaced by $\downarrow$). Hence, $\mathcal{B}_{N,S_z} = \mathcal{B}^{\uparrow}_{N,S_z} \otimes \mathcal{B}^{\downarrow}_{N,S_z}$ has dimension 36. How many states contains the translationally invariant basis $\mathcal{B}^{\uparrow}_{N,S_z,\mathbf{k}}$? To construct this basis, the states of $\mathcal{B}^{\uparrow}_{N,S_z}$ have to be split into groups of states which are mapped onto each other by one of the allowed translations of the cluster. Each group contributes one representative state to $\mathcal{B}^{\uparrow}_{N,S_z,\mathbf{k}}$. The result is (see Fig. 2.5)

$$\mathcal{B}^{\uparrow}_{N,S_z,\mathbf{k}} = \{ |{-}{-}{\uparrow}{\uparrow}\rangle_{(4)}, |{-}{\uparrow}{-}{\uparrow}\rangle_{(2)} \}.$$

(The lower index at each representative state indicates the number N_{diff} of different $\mathcal{B}_{N,S_z}$ states represented by it.)

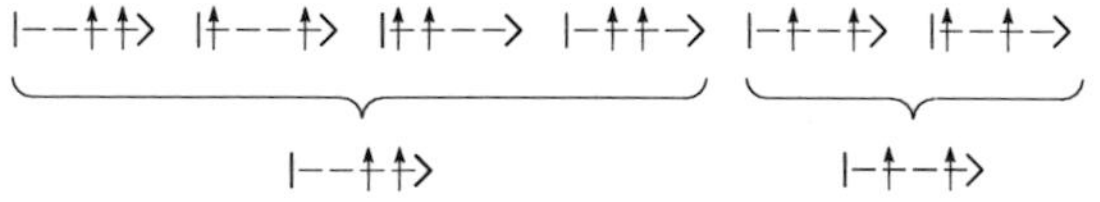

$|{-}{-}{\uparrow}{\uparrow}\rangle$ $|{\uparrow}{-}{-}{\uparrow}\rangle$ $|{\uparrow}{\uparrow}{-}{-}\rangle$ $|{-}{\uparrow}{\uparrow}{-}\rangle$ $|{-}{\uparrow}{-}{\uparrow}\rangle$ $|{\uparrow}{-}{\uparrow}{-}\rangle$

$|{-}{-}{\uparrow}{\uparrow}\rangle$ $|{-}{\uparrow}{-}{\uparrow}\rangle$

Figure 2.5: The states forming the translationally invariant spin-up basis $\mathcal{B}^{\uparrow}_{N,S_z,\mathbf{k}}$, each state being identified by one representative state taken from the basis $\mathcal{B}^{\uparrow}_{N,S_z}$.

The corresponding transition to the momentum eigenbasis for the full basis $\mathcal{B}_{N,S_z}$ yields

$$\begin{aligned}\mathcal{B}_{N,S_z,\mathbf{k}} = \{ \ & |{-}{-}{\updownarrow}{\updownarrow}\rangle_{(4)}, |{-}{\downarrow}{\uparrow}{\updownarrow}\rangle_{(4)}, |{-}{\downarrow}{\updownarrow}{\uparrow}\rangle_{(4)}, |{\downarrow}{-}{\uparrow}{\updownarrow}\rangle_{(4)}, |{\downarrow}{-}{\updownarrow}{\uparrow}\rangle_{(4)}, \\ & |{\downarrow}{\downarrow}{\uparrow}{\uparrow}\rangle_{(4)}, |{-}{\uparrow}{\downarrow}{\updownarrow}\rangle_{(4)}, |{-}{\updownarrow}{-}{\updownarrow}\rangle_{(2)}, |{-}{\updownarrow}{\downarrow}{\uparrow}\rangle_{(4)}, |{\downarrow}{\uparrow}{\downarrow}{\uparrow}\rangle_{(2)} \ \}.\end{aligned}$$

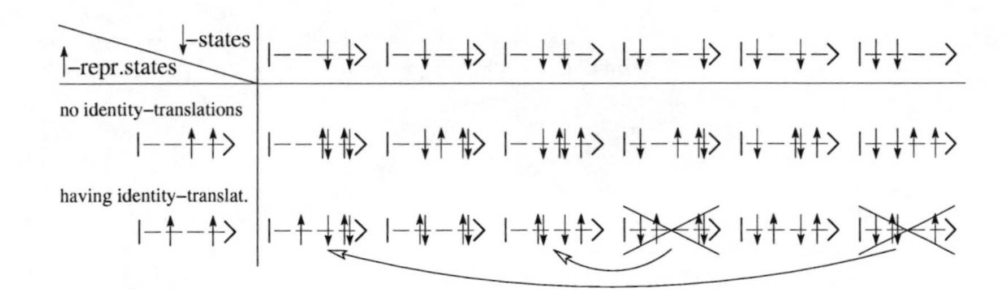

Figure 2.6: Representative states for the translationally invariant basis $\mathcal{B}_{N,S_z,\mathbf{k}}$ can be constructed by combining all translationally invariant spin-up representatives from $\mathcal{B}^{\uparrow}_{N,S_z,\mathbf{k}}$ with all basis states from $\mathcal{B}^{\downarrow}_{N,S_z}$. However, some combinations turn out to be 'false' representatives, i.e. transformations of other representative states (indicated by arrows).

Closer inspection of these 10 states shows that all of them can be constructed by combining one of the two states in $\mathcal{B}^{\uparrow}_{N,S_z,\mathbf{k}}$ with one of the six states in $\mathcal{B}^{\downarrow}_{N,S_z}$ (see Fig. 2.6). This labeling of all states of $\mathcal{B}_{N,S_z,\mathbf{k}}$ by means of $\mathcal{B}^{\uparrow}_{N,S_z,\mathbf{k}}$ and $\mathcal{B}^{\downarrow}_{N,S_z}$ is exactly the type of two-table lookup scheme we were looking for.
However, the table in Fig. 2.6 also contains two 'false' representatives, i.e. combinations of spin-up representative and spin-down basis state which result from certain translations of other combined states in the table (indicated by the two arrows in Fig. 2.6). It is not difficult to show that this can only happen in those rows of the table whose labeling spin-up representatives are states with additional symmetries, i.e. states which represent less than N_o states since they are identically reproduced after less then N_o elementary translations.
As an example, let us determine the nonvanishing matrix element of the hopping term $c^{\dagger}_{1,\uparrow}\, c_{4,\uparrow}$ with $|\,0\,\rangle := |{-}{-}\uparrow\downarrow\uparrow\downarrow\rangle$, the first state in the table in Fig. 2.6, in the $\mathcal{B}_{N,S_z,\mathbf{k}}$ subspace with momentum ($k = \pi$). Our convention is that the orbitals are numbered from left to right, starting with 1; the $\mathcal{B}_{N,S_z,\mathbf{k}}$ basis states are numbered from left to right and from top to bottom in the table in Fig. 2.6, starting with 0. The direct result of applying the hopping term is

$$c^{\dagger}_{1,\uparrow}\, c_{4,\uparrow}\,|{-}{-}\uparrow\downarrow\uparrow\downarrow\rangle = -|\uparrow{-}\uparrow\downarrow\downarrow\rangle, \tag{2.38}$$

where the minus sign is due to the commutation of the moving spin-up electron with the three other electrons in the system. Next, we have to determine the index of the new state in the basis $\mathcal{B}_{N,S_z,\mathbf{k}}$. To this purpose we split up $|\uparrow{-}\uparrow\downarrow\downarrow\rangle$ into the spin-up and the spin-down part, i.e. in $|\uparrow{-}\uparrow{-}\rangle$ and $|{-}{-}\downarrow\downarrow\rangle$. The spin-up state itself is not an element of $\mathcal{B}^{\uparrow}_{N,S_z,\mathbf{k}}$, but it can be obtained from $|{-}\uparrow{-}\uparrow\rangle$ by translation:

$$|\uparrow{-}\uparrow{-}\rangle = \mathcal{T}_{(1)}\,|{-}\uparrow{-}\uparrow\rangle. \tag{2.39}$$

The analogous operation for the spin-down part yields:

$$|{-}{-}\downarrow\downarrow\rangle = -\mathcal{T}_{(1)}\,|\downarrow{-}{-}\downarrow\rangle, \tag{2.40}$$

where the minus sign is again due to the fact that the translation effectively swaps the order of the two spin-down electrons. Combining equations (2.38), (2.39) and (2.40) we get

$$c^{\dagger}_{1,\uparrow}\, c_{4,\uparrow}\, |{-}\,{-}\,{\uparrow\!\downarrow}\,{\uparrow\!\downarrow}\rangle = (-1)^2\, \mathcal{T}_{(1)}\, |\,{\downarrow}\,{\uparrow}\,{-}\,{\uparrow\!\downarrow}\rangle. \tag{2.41}$$

Now, we remember that the state on each side of (2.41) is only a representative (or short cut) for a Bloch state, i.e. a series of translations $\mathcal{T}_{(R)}$ of the representative with phase factors $\exp(ikR)$ (see equation (2.24)). Hence the translation $\mathcal{T}_{(1)}$ on the right hand side of (2.41) has the only effect to generate an overall phase of $\exp(ik1) = -1$ (for our choice of $k = \pi$). In principle, we have completed our task now: the index i of the resulting state is (see Fig. 2.6)

$$\begin{aligned} i &= \mathcal{B}^{\uparrow}_{N,S_z,\mathbf{k}}\text{-index}\big(|{-}\,{\uparrow}\,{-}\,{\uparrow}\rangle\big) \cdot \dim(\mathcal{B}^{\downarrow}_{N,S_z}) + \mathcal{B}^{\downarrow}_{N,S_z}\text{-index}\big(|\,{\downarrow}\,{-}\,{-}\,{\downarrow}\rangle\big) \\ &= 1 \cdot 6 + 3 = 9\,. \end{aligned}$$

Hence, the wanted non-zero matrix element is

$$\langle\,{\downarrow}\,{\uparrow}\,{-}\,{\uparrow\!\downarrow}|\, c^{\dagger}_{1,\uparrow}\, c_{4,\uparrow}|{-}\,{-}\,{\uparrow\!\downarrow}\,{\uparrow\!\downarrow}\rangle = \langle\, 9\,|\, c^{\dagger}_{1,\uparrow}\, c_{4,\uparrow}\,|\,0\,\rangle = (-1)^3 = -1.$$

However, in this special case we have the additional problem that $|\,{\downarrow}\,{\uparrow}\,{-}\,{\uparrow\!\downarrow}\,\rangle$ is one of the rare 'false' representatives in the combined list $\mathcal{B}^{\uparrow}_{N,S_z,\mathbf{k}} \otimes \mathcal{B}^{\downarrow}_{N,S_z}$ shown in Fig. 2.6. Therefore, we have to perform another translation (indicated by an arrow in Fig. 2.6) to find the 'true' representative state $|{-}\,{\uparrow\!\downarrow}\,{\downarrow}\,{\uparrow}\rangle$, which has the combined index $1 \cdot 6 + 2 = 8$. The final result is thus:

$$\langle\, i\,|\, c^{\dagger}_{1,\uparrow}\, c_{4,\uparrow}\,|\,0\,\rangle = \begin{cases} -1\,, & i = 8 \\ \;\;0\,, & i \neq 8. \end{cases} \tag{2.42}$$

The procedure presented above seems relatively complicated. However, it allows to evaluate a whole row of the H matrix with a couple of elementary operations (addition, multiplication, array element access) and with almost no storage requirements: there is no need to store the H matrix, and even creating and storing the Hilbert basis $\mathcal{B}_{N,S_z,\mathbf{k}}$ is not neccessary. A more detailed discussion on efficient implementation, in particular for parallel and/or vectorizing supercomputers, can be found in [65].

t-J-like models

The t-J-model (see section 1.1.2) can be considered as an approximation for the Hubbard model in the case of very large Coulomb repulsion of two electrons in the same orbital ($U \to \infty$). This effectively reduces the Hilbert space by projecting out all states with one or more doubly occupied orbitals. As 't-J-like' models we want to summarize all microscopic many-electron models with the following two properties:

- The model's Hamiltonian conserves the total electron number.
- The model acts on a set of orbitals with three allowed states per orbital: $|\,{-}\,\rangle$, $|\,{\uparrow}\rangle$, and $|\,{\downarrow}\rangle$.

For this type of models, a two-table lookup cannot be implemented by treating the spin-up and the spin-down configurations separately because in the direct product of both partial bases the doubly occupied orbitals inevitably reappear. Much better is a separation of each basis state in a 'charge part' and a 'spin part'. The procedure shall be demonstrated at the example of a 4-orbital chain with two spin-up electrons and one spin-down electrons; three orbitals ('sites') are occupied by exactly one electron and one site is empty. In this case, the Hilbert space basis $\mathcal{B}_{N,S_z}$ has the dimension $4{\cdot}3 = 12$ since there are 4 possibilities to choose the empty site and for each of these configurations there are 3 possibilities to place the spin-down electron on one of the three non-empty sites.
An efficient way of coding the basis states is the following: use $N_o = 4$ bits to mark the free ($\rightarrow$0) and occupied ($\rightarrow$1) orbitals and use $N_{el} = 3$ bits to store the z-spin direction of each occupied orbital (0 for spin-down, 1 for spin-up). The state $|\uparrow - \downarrow \uparrow\rangle$, for example, is then coded as

$$|\uparrow - \downarrow \uparrow\rangle \rightarrow |1011;101\rangle\,.$$

Calling the first N_o bits the charge bits and the last N_{el} bits the spin bits, we see that all 12 states of $\mathcal{B}_{N,S_z}$ can be written as a direct product of the charge subbasis $\mathcal{B}^{ch}_{N,S_z}$ and the spin subbasis $\mathcal{B}^{sp}_{N,S_z}$:

$$\mathcal{B}_{N,S_z} = \mathcal{B}^{ch}_{N,S_z} \otimes \mathcal{B}^{sp}_{N,S_z}$$

with

$$\begin{aligned} \mathcal{B}^{ch}_{N,S_z} &= \{\, |0111\rangle, |1011\rangle, |1101\rangle, |1110\rangle \,\} \quad \text{and} \\ \mathcal{B}^{sp}_{N,S_z} &= \{\, |011\rangle, |101\rangle, |110\rangle \,\}\,. \end{aligned}$$

Details on exploiting further symmetries within this approach can be found in [82].

Heisenberg-like models

The spin-1/2 Heisenberg model [43] corresponds to the t-J model at half-filling, i.e. if each orbital is filled with exactly one electron. As 'Heisenberg-like' models we understand all many-electron models describing the magnetic interactions of n spin-1/2 particles which are pinned at n different lattice sites. This implies that there are two allowed states at each site: $|\uparrow\rangle$ and $|\downarrow\rangle$.
The best way to set up a two-table lookup scheme for Heisenberg-like models is an 'alternating site' or 'checkerboard' splitting of the lattice in two sublattices [82]. The 8-site chain $|\uparrow\uparrow\downarrow\uparrow\uparrow\downarrow\downarrow\downarrow\rangle$, for example, could be split into 4 bits describing the even sites and 4 bits describing the odd sites of the lattice.

2.3.5 Methods for single molecules or metal-ligand clusters

The difficulties and challenges in applying the Lanczos method to single molecules or metal-ligand clusters are completely different from those encoutered when dealing with highly ordered periodic lattices. In the latter case, the main challenges have been to reduce the dimension of Hilbert space using the symmetries of Hamiltonian and lattice and to implement a fast matrix-free access to the elements of the H matrix. In the case of single molecules or metal-ligand clusters with open boundary conditions (OBC), it is very often not worth spending much effort on exploiting symmetries: translational invariance is out of the game due to the OBC, and if the cluster is distorted or the molecule not highly symmetric, then there are not even any point group symmetries to be exploited. Furthermore, as we will see below, recurring to matrix-free methods can often be avoided. On the other hand, there are new challenges, one of them being the demand for high precision results: experiments on molecules or metal-ligand clusters, in particular optical or X-ray absorption or photoemission experiments, often produce high precision data with relative errors of 10^{-4} or better; these experiments may reveal fine structure effects like spin-orbit coupling [83] which must be accounted for by the numerical simulation in order to describe the experiments correctly. When studying microscopic models like the Hubbard model on periodic lattices, on the contrary, high precision is rarely required. Rather, one wants to to know whether or not these models show certain qualitative properties such as AF ordering or a superconducting phase. Another complication when simulating optical absoprtion or emission experiments might be the excitation of core electrons, i.e. the total number of orbitals to take into account might vary during the simulation.

In the following, we discuss the resulting implementation problems at the example of a code for the simulation of X-ray spectroscopy on metal-ligand clusters with one central metal atom and several ligand orbitals. More precisely, we are going to model a system of

- $2L+1$ valence orbitals of a metal atom, labeled by their orbital angular momentum in z-direction $L_z = -L, ..., L$ (L is the total orbital angular momentum, i.e. $L = 1$ for p metals, $L = 2$ for d metals (such as the transition metals), $L = 3$ for $4f$ or $5f$ metals).

- a certain number of bonding linear combinations of ligand orbitals – e.g. of $2p$ orbitals of oxygen ligands – with zero orbital angular momentum.

- $2(L-1)+1$ core orbitals of the metal atom with total orbital angular momentum $L-1$, thus permitting excitations of core electrons into the valence shell which fulfill optical selection rules [83]. In the ground state of the system this shell is completely filled, in the optically excited state it carries one hole.

In order to describe real optical or X-ray experiments such as X-ray magnetic circular dichroism (see section 2.4.2), we study Hamiltonians with the following energetic interactions

- a different orbital energy offset for each of above orbitals

- Coulomb repulsion and electron-electron exchange interaction within the metal's valence and core orbitals (Slater-Koster theory, see [84]).

- potential interaction between core holes and the metal's valence electrons.

- single electron hopping between different ligand orbitals and between ligand orbitals and the valence orbitals of the metal atom.

- electron pair hopping conserving L_z on the metal's valence orbitals.
- anisotropic crystal field potentials modeling arbitrary distorted cluster geometries. Another effect of crystal field is to induce single electron hopping on the metal's valence shell.
- spin orbit coupling on the valence and the core orbitals of the metal atom.
- L_z exchange and spin exchange interaction between electrons in the metal's core and valence shell.
- Magnetic interaction with an external magnetic field; the field can have an arbitrary angle with the cluster's main axes and couples to the electrons' orbital angular momenta and spins with two different effective g-factors [83].

For the theoretical motivation of these interaction terms, we refer to [84] and to textbooks on the physics of atoms and molecules, e.g. [83]. When trying to reproduce experimental X-ray dichroism spectra (see section 2.4.2), we found that each of these terms is needed to get satisfying results.
As to the implementation of the Lanczos method, the first idea might be to split up the full basis into a spin-up and a spin-down part and to implement a two table lookup just as it was done in section 2.3.4 for the Hubbard model. However, spin-orbit coupling does not permit this approach here because it induces spin flips, and thus prevents the numbers of spin-up and spin-down electrons, $n_\uparrow$ and $n_\downarrow$, from being constant. Another consequence of the spin flips is that the Hilbert space dimension is inflated by a factor of 3 to 4 (see the table at the beginning of this chapter) because $n_\uparrow$ and $n_\downarrow$ cannot be kept constant any more, only the total electron number n_{el} remains constant.
In this situation it is most convenient to organize the bookkeeping of the Hilbert basis states in 4 levels (see Figures 2.7 and 2.8):

- The innermost level runs through all core states. In the ground state, this level contains only one state, the completely filled core (see Fig. 2.7); after X-ray absorption, there are $2(2L-1)$ different core states, labeled by the spin orientation and L_z value of the missing electron (see Fig. 2.8).
- The next level runs through all ligand states with a fixed total number of electrons in all ligand orbitals n_{lig}.
- The second outermost level analogously runs through all electron configurations of the metal's valence shell with fixed total number of valence electrons $n_{val} = n_{el} - n_{lig}$.
- The outermost level runs through all possible distributions of the electrons on ligand and metal orbitals, i.e. through all possible values of n_{lig} and $n_{val} = n_{el} - n_{lig}$.

This organization of the Hilbert basis has two major advantages. First, it assures that all energetic interactions given above either apply only to small submatrices of the H matrix which are identically reproduced many times within H (e.g. ligand-ligand hopping), or they create long sequences of identical matrix elements parallel to the main diagonal of H (e.g. hopping, pair hopping and spin-orbit coupling processes within the valence metal's

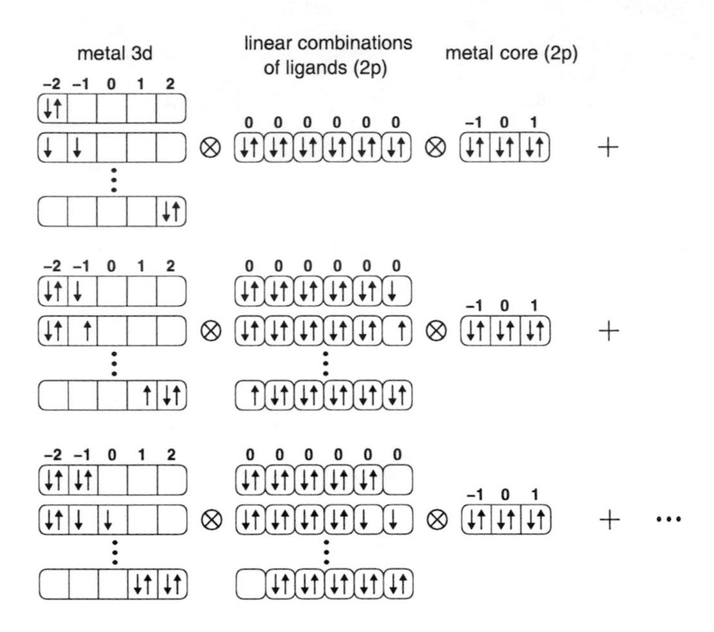

Figure 2.7: The basis states of a metal-ligand cluster system can be characterized by the electron distribution between ligand orbitals and the metal's valence shell, and for fixed such distribution by the distribution of the electrons onto the different valence shell and ligand orbitals. The situation printed here describes a transition metal atom (with 3d valence shell) and oxygen ligands (with 2p valence orbitals). The numbers above the orbitals are orbital angular momentum numbers L_z.

shell). Both structures allow to store the H matrix in a very compressed form: each type of small submatrices needs to be stored only once, and for each diagonal sequence of identical elements one only has to store position and value of the first element and the length of the sequence. For this reason, one can easily store the H matrix even for Hilbert space dimensions of largely above 10^6, which corresponds to 12 to 15 orbitals. This is enough to model all practically relevant cluster geometries with a p-, d- or f-metal in the center. Furthermore, the existence of many sequences of identical matrix elements parallel to the diagonal of the matrix is best suited for speeding up the computation of $|y\rangle = H|x\rangle$ since in this situation the elements of $|y\rangle$ *and* $|x\rangle$ can be traversed step by step in correct storage order. This is the ideal situation for pipeline prediction in modern processor architectures.

The second advantage of organizing the Hilbert basis in the four layers described above is that it permits a systematic approximation which projects out 'irrelevant parts' of Hilbert space and thereby reduces the dimension of the H matrix by factors of 2 to 100. To understand this, one has to keep in mind that in almost all real world metal-ligand clusters the ligands have a much stronger electronegativity than the central metal atom.

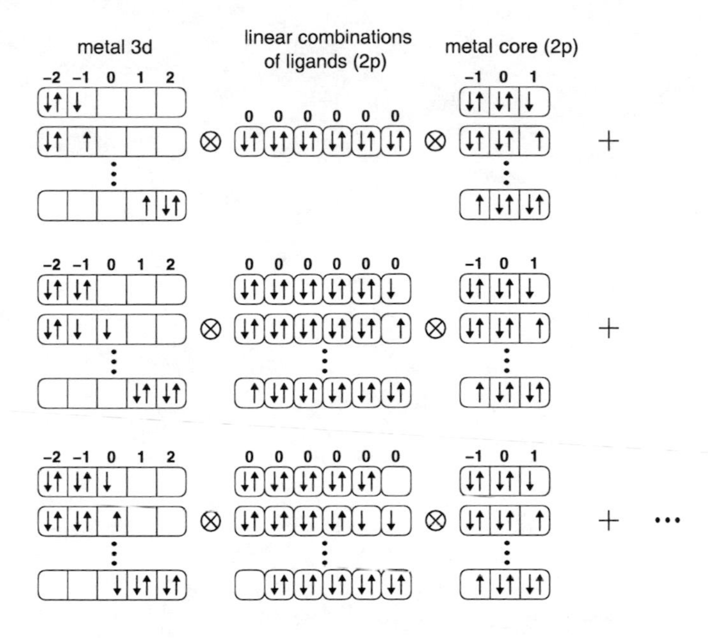

Figure 2.8: The basis states of a metal-ligand cluster after a core electron has been excited into the metal's valence shell by X-ray absorption.

In the case of CrO_2, for example, in which each chromium atom is surrounded by an octahedron of 6 oxygen atoms, the oxygen atoms tend to attract two more electrons and to reach a saturated $1s^2 2s^2 2p^6$ configuration; in this scenario, the chromium atom looses four of its electrons and ends up in an $[\mathrm{Ar}]3d^2$ configuration. If we run an ED simulation of such a cluster of 6 oxygen orbitals and 5 chromium $3d$ orbitals, we therefore expect that the ground state of the system will consist of a large contribution of $3d^2$ states and some slightly mixed in $3d^3$ and maybe $3d^4$ contributions. All states with more than 4 electrons in the chromium's valence shell are unlikely to contribute to the cluster's low energy physics and can probably be omitted when constructing the system's Hilbert space. Indeed, ED calculations with realistic interaction parameters for CrO_2 show that the ground state contains about 95% $3d^2$ contributions, about 4.5% $3d^3$ contributions, 0.5% $3d^4$ and less than 0.02% of other configurations. This means that projecting out all higher configurations than $3d^4$ does only introduce a relative truncation error of about 10^{-4}. Table 2.2 shows that this approximation reduces the dimension of the H matrix by a factor of about 96% in the ground state and by 88% in the state with core electron excitation.

In appendix A.2 we document a computer program (in C++) which implements the ideas described here. Section 2.4.2 presents an exemplary application of this program: the interpretation of circular magnetic X-ray dichroism spectra [13] of the technologically interesting material CrO_2.

configuration	dim_{gs}	dim_{excit}
$3d^2$	45	720
$3d^3$	1440	15120
$3d^4$	13860	99792
⋮	⋮	⋮
total	319770	1023264

Table 2.2: Hilbert space of an octahedric cluster consisting of one chromium atom (valence shell of 5 3d orbitals) surrounded by 6 oxygen 2p orbitals. The left column contains the number of electrons in the 3d valence shell, dim_{gs} is the Hilbert space dimension before X-ray excitation of a 2p core electron, dim_{excit} the dimension after creating such an excitation.

2.4 Exemplary applications

2.4.1 Metal-insulator transition in the Hubbard model

This section presents a typical application of ED methods to highly ordererd periodic crystal lattices: we study the metal-insulator transition in the half-filled Hubbard model [11] with additional second-nearest neighbor hopping terms. Parts of the results to be presented in this section can be found in [12]. This publication also summarizes some important dynamical mean field theory (DMFT) results [85–88] on the Hubbard model in $d = \infty$ dimensions.

In section 1.1.1 we have seen that theoretical considerations and earlier numerical simulations [4, 34, 35] of the Hubbard model predict an interesting competition between two different insulating states – the AF ordered Mott-Heisenberg insulator and the Mott-Hubbard insulator due to electron-electron repulsion – and a metallic state. This scenario raises two questions: is there a critical value U_c at which the system undergoes a sudden metal-insulator transition (MIT)? And of which type is the insulating state, Mott-Hubbard or Mott-Heisenberg? Our expectation is that we find a MIT of Mott-Heisenberg type at $T = 0$ and one of Mott-Hubbard type at finite temperature because the Mermin-Wagner theorem [35] prohibits long range AF order in two-dimensional systems. However, when running numerical simulations on finite clusters, the system already 'sees' overall AF order as soon as the staggered spin-spin correlation length is of the same order as the cluster length. Therefore any insulating state detected on the finite cluster has to be checked carefully in order to decide whether it is a true property of the infinite lattice or a finite-size artefact.

We have attacked above questions by means of two different numerical techniques: quantum Monte Carlo simulations (QMC) at finite temperature $T > 0$ and Exact Diagonalization at $T = 0$. In order to suppress the Mott-Heisenberg insulating state even at $T = 0$ and to reduce local AF correlations at $T > 0$, we have added an additional hopping term t' acting between (diagonal) second nearest neighbors. The studied Hamiltonian thus reads

$$H = U \sum_i n_{i,\uparrow} n_{i,\downarrow} - t \sum_{\langle i,j \rangle,\sigma} (c^\dagger_{i,\sigma} c_{j,\sigma} + \text{h.c.}) - t' \sum_{\langle\langle i,j \rangle\rangle,\sigma} (c^\dagger_{i,\sigma} c_{j,\sigma} + \text{h.c.}) , \tag{2.43}$$

where $\langle i,j \rangle$ denotes all pairs of nearest neighbor sites and $\langle\langle i,j \rangle\rangle$ all pairs of second nearest

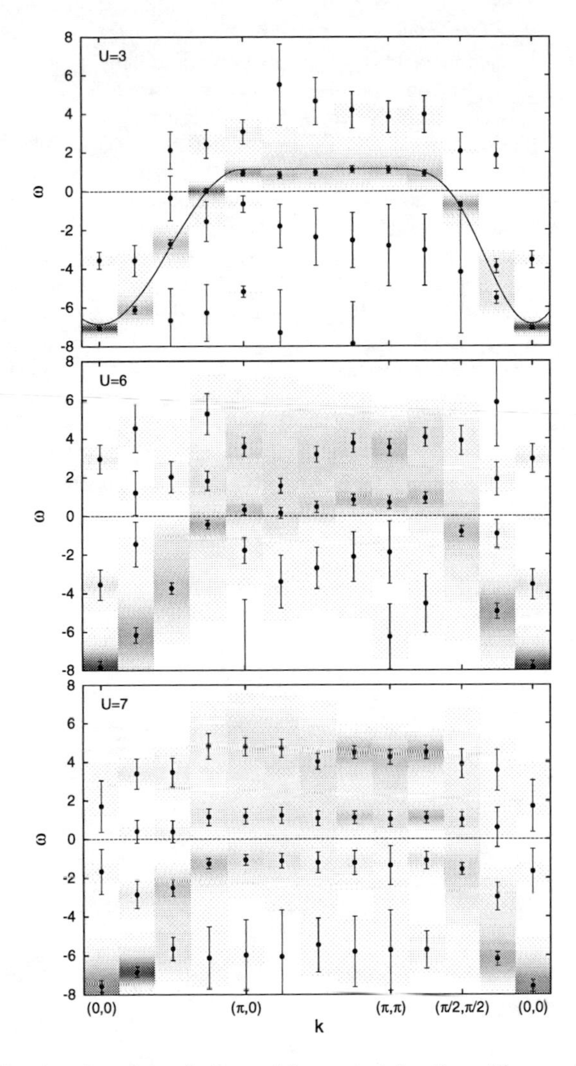

Figure 2.9: Density plot of the single particle spectral function with second nearest neighbor hopping t' for an 8×8 lattice with $\beta t = 3$, $U/t = 3$ (top), 6 (middle) or 7 (bottom), and $t'/t = 0.5$ as obtained by QMC. At $U/t = 6$ a gapless band crossing the Fermi energy at $\mathbf{k} = (\pi, 0)$ and $\mathbf{k} = (\pi/2, \pi/2)$ is still visible. Between $U/t = 6$ and $U/t = 7$, a metal-insulator transition is happening. The insulating state is of the Mott-Hubbard type due to the absence of reflected structures (QMC result by M.G. Zacher).

neighbors. If t' and t are both positive, the AF ordered phase is strongly disfavored by frustration. In our numerical simulations we chose $t' = 0.5\,t$ and $t' = t$.

Figure 2.9 shows the single particle spectral function (describing photoemission and inverse photoemission of an electron) obtained by QMC simulations at $T/t = 1/3$. Between $U/t = 6$ and $U/t = 7$, a metal-insulator transition is clearly visible. Two observations tell us that this MIT is indeed of Mott-Hubbard type. First, there are no reflected structures in the spectral function for $U = 7$. And second, an analysis of the staggered spin-spin correlation yields a correlation length of about 1.5, which is considerably smaller than the lattice extension of 8 (for more details see Ref. [12]).

Now we study the $T = 0$ case by Exact Diagonalization of a 4×4 lattice, discussing ED results for $t'/t=0$, 0.5 and 1, and for U/t varying from 2 to 11. Three different kinds of results will be presented. First, ED allows to calculate the quantum numbers of energy E, the square of total spin $S(S+1)$ and momentum $\mathbf{p}=\hbar\mathbf{k}$ of the ground state. Second, as shown in section 2.2.3, ED can be used to calculate single-particle dynamical response functions $A(\mathbf{k},\omega)$. From these functions one can derive the Fermi energy and the band structure of the system, just as we have done it in Fig. 2.9 for the QMC data. Third, ED also finds the exact quantum mechanical ground state $|\,\mathrm{gs}\,\rangle$ of the system itself. Therefore we can check directly up to which extent the system is antiferomagnetically ordered and whether the electrons are strongly localized or heavily fluctuating. To this purpose, we ask which basis states $|\,b\,\rangle$ – we work in the $\mathcal{B}_{N,S_z,\mathbf{k}}$ basis, see section 2.3.2 – have the strongest overlap integral $\langle b|\mathrm{gs}\rangle$ with the ground state. Figure 2.10 shows representatives for the three basis states which can be expected to be most important in an AF ordered system: the perfectly ordered AF state $|\,\mathrm{AF}_0\,\rangle$ and its smallest possible spin and charge excitations $|\,\mathrm{AF}_1\,\rangle$ and $|\,\mathrm{AF}_2\,\rangle$. If the overlap of these states with $|\,\mathrm{gs}\,\rangle$ is large, then the system is strongly AF ordered.

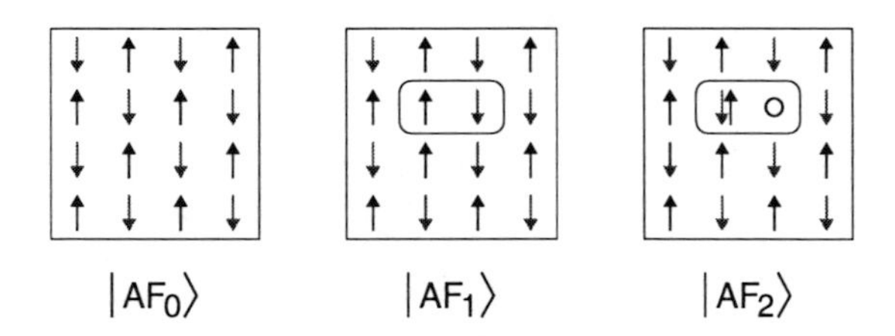

Figure 2.10: The perfectly AF ordered state of a 4×4 cluster (left) and its first spin (middle) and charge (right) excitations

The tools being described, we proceed to the ED results themselves. To test our theoretical predictions on the AF character of the ground state at $t'=0$, we first calculate $A(\mathbf{k},\omega)$ at $U/t=6$ and $t'=0$ (see Fig. 2.11). The system possesses an energy gap of about $3.2\,t$, and the reflected band structures originating from AF ordering and Brillouin zone shrinking (compare Fig. 1.3) are clearly visible. Thus, the system is a Mott-Heisenberg insulator as predicted by the literature [34].

The overlap integrals $\langle\mathrm{AF}_{i=0,1,2}|\mathrm{gs}\rangle$ for the Hubbard model with $t'=0$ are

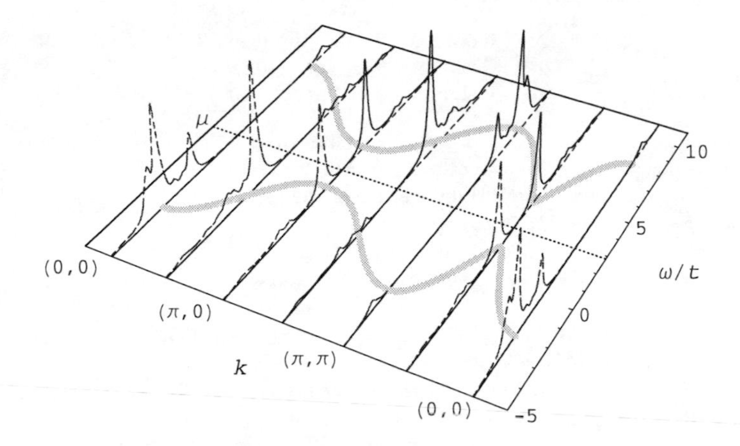

Figure 2.11: *single particle spectral function $A(\mathbf{k},\omega)$ for a 4×4 lattice with $U/t=6$ and $t'=0$, obtained by ED. μ is the Fermi energy; the dashed peaks below μ are photoemission peaks and correspond to occupied energy bands in the infinite lattice, the solid peaks above μ are inverse photoemission peaks and correspond to empty bands in the infinite system. The reflected band structures indicate that the system is in a Mott-Heisenberg insulating state.*

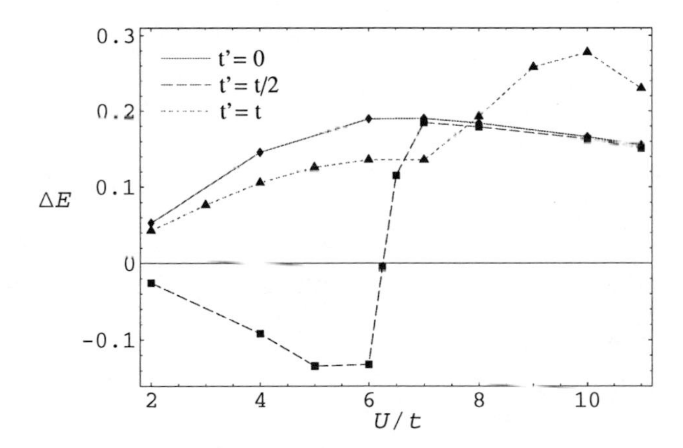

Figure 2.12: *Energy difference between the lowest eigenenergies of the $\mathbf{k}=(0,0)$, $S=0$ and the $\mathbf{k}=(\pi,\pi)$, $S=1$ subspaces of Hilbert space: $\Delta E = E(\mathbf{k}=(\pi,\pi)) - E(\mathbf{k}=(0,0))$. For $\Delta E > 0$, the $\mathbf{k}=(0,0)$ state is the true ground state state of the system, for $\Delta E < 0$ the gound state has $\mathbf{k}=(\pi,\pi)$ and $S=1$. The diamonds connected by a solid line represent the Hubbard model without second nearest neighbor hopping ($t'=0$), the squares connected by a dashed line show the model with $t'=t/2$ and the triangles connected by a dotted line the one with $t'=t$.*

U/t	$\langle \mathrm{AF}_0 \vert \mathrm{gs}\rangle$	$\langle \mathrm{AF}_1 \vert \mathrm{gs}\rangle$	$\langle \mathrm{AF}_2 \vert \mathrm{gs}\rangle$
2	0.06125	0.09754	0.06397
6	0.20350	0.21435	0.18980
10	0.30583	0.33146	0.19001

In all three cases, $U/t=2$, 6, and 10, the three states $|\,\mathrm{AF}_{i=0,1,2}\rangle$ have significantly larger overlap integrals with $|\,\mathrm{gs}\rangle$ than all other states. (The average value of $\langle\, b\,|\mathrm{gs}\rangle$ for a random basis state $|\,b\,\rangle$ is 0.00086 as the Hilbert space of the half filled 4×4 Hubbard model has dimension $dim \approx 1350000$ if all translational and point symmetries are exploited [79, 80]). In summary, our results again indicate that the Hubbard model without second nearest neighbor hopping is a Mott-Heisenberg insulator for all nonvanishing U (at least for $U \gtrsim 1$).
Now we switch on t'. When calculating the E, $S(S+1)$ and $\mathbf{k}$ expectation values of the ground state, we make a striking observation: for $t'=0$ and $t'=t$, $|\,\mathrm{gs}\,\rangle$ has $S=0$ and $\mathbf{k}=(0,0)$ on the entire studied U range; for $t'=t/2$, however, this is only true for $U>U_c:=6.25\,t$. At U_c, a level crossing takes place, and below this value the ground state has $S=1$ and $\mathbf{k}=(\pi,\pi)$. This is shown in Fig. 2.12, in which the energy difference

$$\Delta E := E(\mathbf{k}{=}(\pi,\pi)) - E(\mathbf{k}{=}(0,0))$$

is plotted. $E(\mathbf{k}{=}(0,0))$ and $E(\mathbf{k}{=}(\pi,\pi))$ are the ground state energies in the subspaces with $\mathbf{k}=(0,0)$ and $\mathbf{k}=(\pi,\pi)$, respectively.
The almost identical $\Delta E(U)$ lines for $t'=0$ and $t'=t/2$ above U_c suggest that the $t'=t/2$ system might still be a Mott-Heisenberg insulator at $U>U_c$. But what is the character of the $t'=t/2$ ground state below U_c? And does the sharp kink of the $t'=t$ line at $U\approx 7$ indicate that the $t'=t$ system also undergoes a certain phase transition at a U value slightly above U_c? To answer these questions, we have to analyse the ground states and the $A(\mathbf{k},\omega)$ spectra below and above $U/t=6.25$ in the case $t'=t/2$, and below and above $U/t\approx 7$ for $t'=t$. Interestingly, both 'critical' U values are very close to the value at which the metal-insulator transition has been detected in QMC at finite temperatures.
A comparison of the $t'=t/2$ and the $t'=0$ AF overlap integrals yields

	$t'=t/2$			$t'=0$		
U/t	$\langle \mathrm{AF}_0 \vert \mathrm{gs}\rangle$	$\langle \mathrm{AF}_1 \vert \mathrm{gs}\rangle$	$\langle \mathrm{AF}_2 \vert \mathrm{gs}\rangle$	$\langle \mathrm{AF}_0 \vert \mathrm{gs}\rangle$	$\langle \mathrm{AF}_1 \vert \mathrm{gs}\rangle$	$\langle \mathrm{AF}_2 \vert \mathrm{gs}\rangle$
6	0.05591	0.09388	0.08366	0.31167	0.27147	0.28547
7	0.22654	0.25749	0.18974	0.23679	0.25122	0.19696

Here, the data for $U/t=6$, $t'=0$ are not given for the ground state (which has $\mathbf{k}=(0,0)$ and $S=0$) but for the lowest energy eigenstate of the $\mathbf{k}=(\pi,\pi)$ and $S=1$ subspace. It can be clearly seen that above U_c the AF order remains almost unchanged after activating $t'=t/2$, whereas it is strongly suppressed in the new ($\mathbf{k}=(\pi,\pi)$, $S=0$) ground state below U_c. This suggests that the Mott-Heisenberg insulator is indeed suppressed at $U<U_c$ in the $t'=t/2$ case.
Stronger evidence for the presumed phase transition at U_c can be obtained from the $A(\mathbf{k},\omega)$ spectra plotted in Fig. 2.13. At $U/t=7$, the signatures of the Mott-Heisenberg insulator – energy gap and reflected band structures – remain almost unchanged with respect to the $S=0$ spectrum in Fig. 2.11. The spectrum at $U/t=6$, however, bears

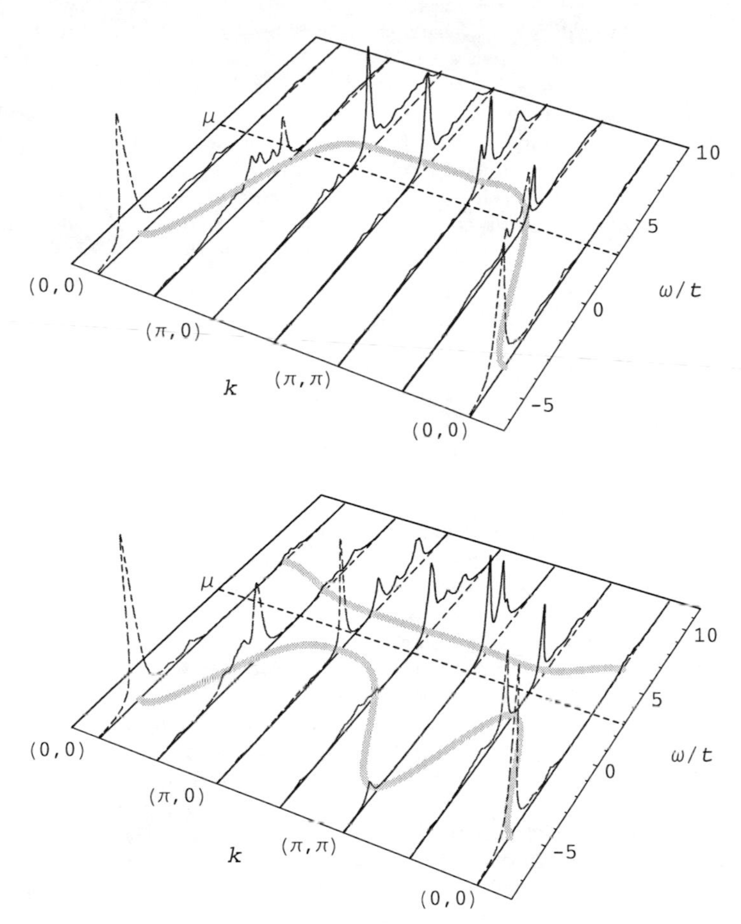

Figure 2.13: single particle spectral functions $A(\mathbf{k},\omega)$ for a 4×4 lattice with $t'=t/2$ and $U/t=6$ (top) resp. $U/t=7$ (bottom), obtained by ED. μ is the Fermi energy; the dashed peaks below μ are photoemission peaks and correspond to occupied energy bands in the infinite lattice, the solid peaks above μ are inverse photoemission peaks and correspond to empty bands in the infinite system. For $U/t=6$, the system is metallic with one single band crossing the Fermi energy. At $U/t=7$, however, a gap opens, and the system shows the reflected band structures of a Mott-Heisenberg insulator.

a strong similarity with the corresponding finite temperature QMC spectrum given in Fig. 2.9 and has a clearly metallic character. In particular, the gap at $\mathbf{k}=(\pi/2,\pi/2)$ is completely closed.

In summary, we have identified a sharp metal-insulator transition of Mott-Heisenberg type at $U_c = 6.25\,t$ in the system with $t' = t/2$. This transition coincides with a level crossing and a change in the spin and momentum quantum numbers of the ground state. We have seen that at $T=0$ the second nearest neighbor hopping $t'=t/2$ is not sufficient to completely destroy AF order at large U values – in contrast to the finite temperature results obtained by QMC. If we want to drive the system into an MIT of Mott-Hubbard type, we have to further increase t'. A value of $t'=t$ seems best suited for this purpose as it guarantees 'maximum frustration' of the two hopping terms t and t'. Indeed, the AF overlap integrals now signalize the breakdown of AF order in the entire studied U range:

U/t	$\langle \mathrm{AF}_0 \vert \mathrm{gs}\rangle$	$\langle \mathrm{AF}_1 \vert \mathrm{gs}\rangle$	$\langle \mathrm{AF}_2 \vert \mathrm{gs}\rangle$
2	< 0.01	< 0.01	< 0.01
7	< 0.01	0.0854	< 0.01
11	< 0.02	< 0.06	< 0.02

(In above table, an upper boundary instead of an exact value means that the corresponding state $|\mathrm{AF}_i\rangle$ is not one of the 20 basis states with largest overlap with $|\mathrm{gs}\rangle$.)

The $A(\mathbf{k},\omega)$ spectra again show the signatures of a metal at small U values (Fig. 2.14 top). A side effect of the strong second nearest neighbor hopping t' is to push down the peak at $\mathbf{k}=(\pi,\pi)$ below the Fermi energy – a tendency that could be expected from a comparison of the corresponding metallic states at $t'=0$ (Fig. 2.11) and $t'=t/2$ (Fig. 2.13 top). At the large Coulomb interaction of $U/t=11$, however, the single particle spectrum is very similar to the finite temperature spectrum found at $U/t = 7$ and $t' = t/2$ in QMC (compare Figures 2.9 and 2.14 bottom): there are four different energy bands with only small dispersion The system is thus a Mott-Hubbard insulator. Now the question is whether a sharp phase transition between metal and Mott-Hubbard insulator can be detected. The ground state energy diagram in Fig. 2.12 shows a sharp kink between $U/t=7$ and 7.5 for $t'=t$. So this range of U values is our first candidate for the transition to take place. Indeed, the single particle spectrum for $U/t=6$ in Fig. 2.15 is still of more metallic character with one single band crossing the Fermi edge, while in the spectrum at $U/t=9$ a considerable amount of peak weight remains below (respectively above) the Fermi energy for all $\mathbf{k}$ values, and the four bands of the Mott-Hubbard insulator appear. However, a more precise determination of the critical U value can not be deduced from the ED spectra. This is a typical consequence of finite-size effects as we have discussed them in section 1.3: the smaller a system is, the broader becomes the transition regime around a phase transition. It should also be noted that we have classified the $U/t=6$ spectrum as 'metallic' because of its general shape even though it has a finite gap at $\mathbf{k}=(\pi/2,\pi/2)$. This is good example for what we have stated in section 1.3: that small finite gaps in ED spectra of small clusters must not be taken too seriously.

Altogether, our ED studies on metal-insulator transitions in the Hubbard model at zero temperature can be summarized in the phase diagram shown in Fig. 2.16: as a function of Coulomb repulsion U/t and second nearest neighbor hopping t'/t, there are two insulating phases I_1 (Mott-Heisenberg insulator) and I_2 (Mott-Hubbard insulator) and a metallic

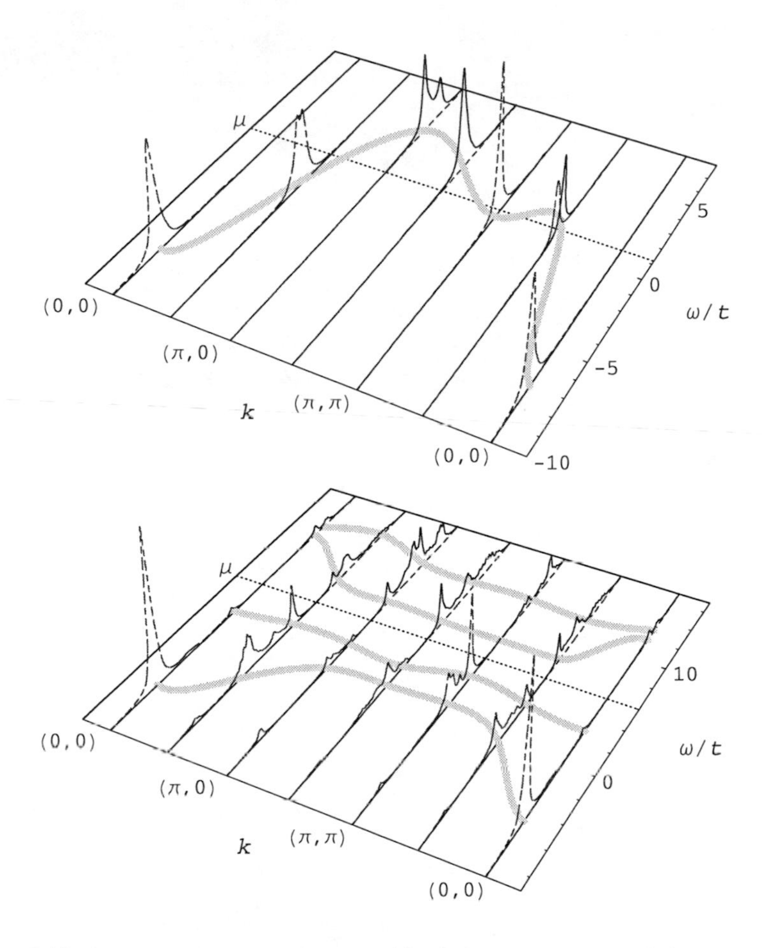

Figure 2.14: single particle spectral functions $A(\mathbf{k},\omega)$ for a 4×4 lattice with $t'=t$ and $U/t=2$ (top) resp. $U/t=11$ (bottom), obtained by ED. μ is the Fermi energy; the dashed peaks below μ are photoemission peaks and correspond to occupied energy bands in the infinite lattice, the solid peaks above μ are inverse photoemission peaks and correspond to empty bands in the infinite system. For $U/t=2$, the system is metallic with one single band crossing the Fermi energy. The large value of t' has the effect to push down the peak at $\mathbf{k}=(\pi,\pi)$ below the Fermi energy (compare Figures 2.11 and 2.13 top). At $U/t=11$, however, a gap opens, and the system shows the four bands which are also present in the QMC results at finite temperature. The absence of reflected band structures indicates that the system is in an insulating state of Mott-Hubbard type.

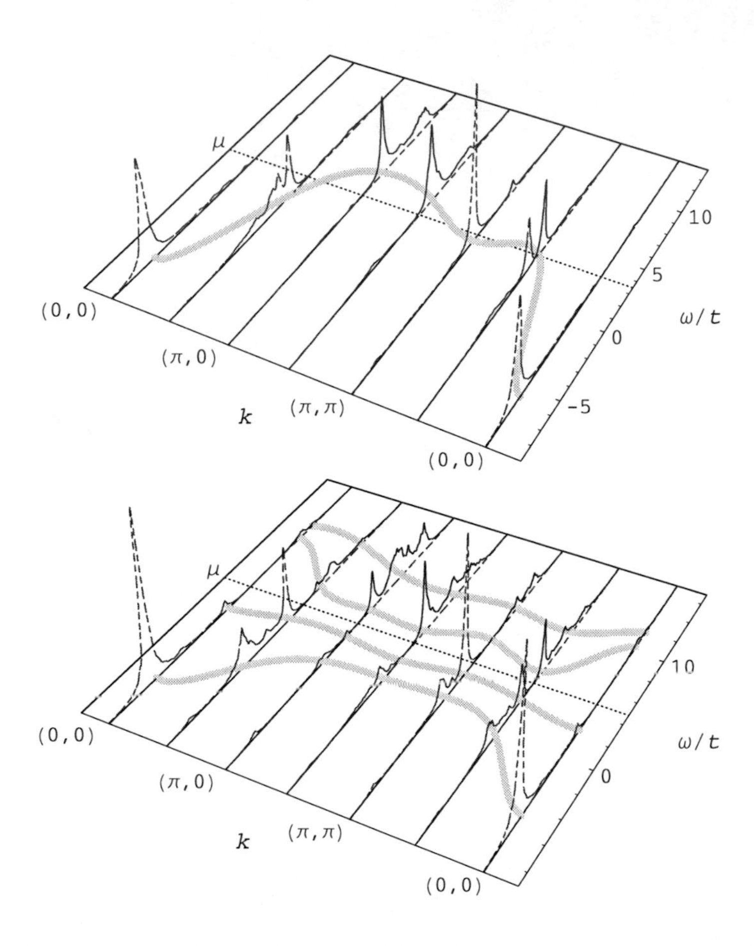

Figure 2.15: single particle spectral functions $A(\mathbf{k},\omega)$ for a 4×4 lattice with $t'=t$ and $U/t=6$ (top), resp. $U/t=9$ (bottom), obtained by ED. μ is the Fermi energy; the dashed peaks below μ are photoemission peaks and correspond to occupied energy bands in the infinite lattice, the solid peaks above μ are inverse photoemission peaks and correspond to empty bands in the infinite system. The plots show that the phase transition from the metal to the Mott-Hubbard insulator takes place somewhere between $U/t=6$ and $U/t=9$, i.e. at the value $U\approx7$ expected from Fig. 2.12.

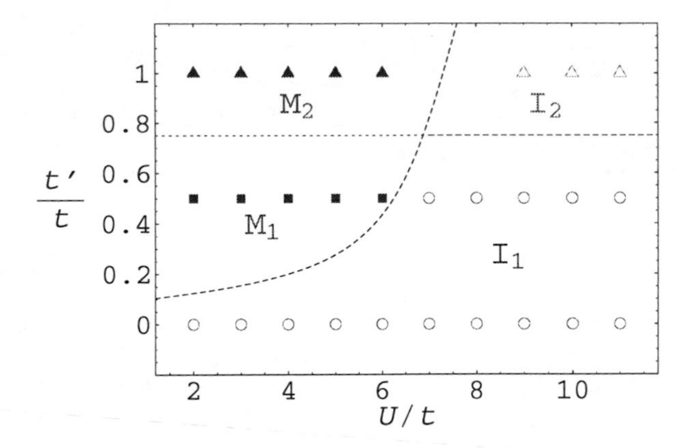

Figure 2.16: The $T=0$ phase diagram of the 2D Hubbard model as a function of Coulomb repulsion U/t and second nearest neighbor hopping t'/t as it is suggested by the ED results presented in this section. There are two insulating phases I_1 (Mott-Heisenberg insulator) and I_2 (Mott-Hubbard insulator) and a metallic phase with two different types of ground states: the $\mathbf{k}=(\pi,\pi)$, $S=1$ state at small values of t' (M_1 phase) and the $\mathbf{k}=(0,0)$, $S=0$ state at larger t' (M_2 phase).

phase with two different types of ground states: the $\mathbf{k}=(\pi,\pi)$, $S=1$ state at small values of t' (M_1 phase) and the $\mathbf{k}=(0,0)$, $S=0$ state at larger t' (M_2 phase).

2.4.2 X-ray magnetic circular dichroism in CrO_2

X-ray magnetic circular dichroism (XMCD) is a relatively new experimental technique for studying magnetic material properties [13]. XMCD exploits the different absorptions of left and right circular polarized X-rays in most magnetic materials which is due to different densities of states for spin-up and spin-down electrons at the Fermi edge. This effect is caused by magnetic exchange interactions [5] between the valence electrons of magnetic materials: the magnetic exchange creates an effective internal magnetic field which induces an energy offset between spin-up and spin-down electrons; this shifts the density of states curve of spin-up electrons with respect to the corresponding curve for the spin-down electrons, so that the two densities of state at the Fermi edge may differ considerably (see Fig. 2.17). The XMCD spectrum is the difference between the two X-ray absorption spectra (XAS) for left and right circular polarized X-rays. It contains informations on the densities of states above the Fermi edge and permits to determine the spin and orbital magnetic moments of the ground state.

Chromium dioxide is a material with several interesting properties for science and technology [89]. It shows an almost perfect spin polarization at the Fermi edge and has semimetallic properties, therefore it is a promising system for the development of spin-electronic devices. For physicists, the material is interesting because of the combination

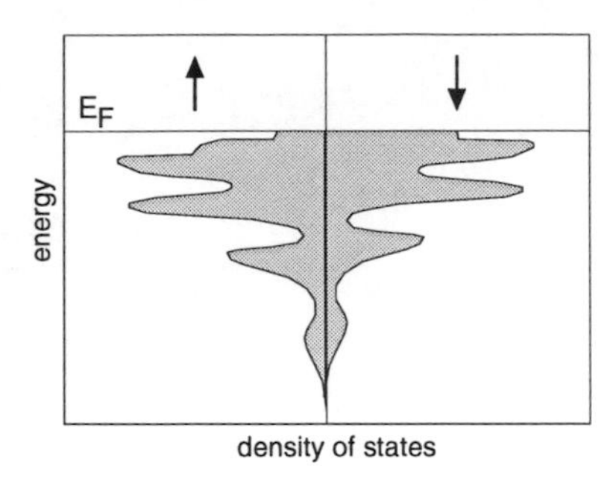

Figure 2.17: The magnetic exchange interaction between valence electrons in magnetic materials acts like an internal magnetic field, which induces an offset in the density of states spectra between spin-up and spin-down electrons. Therefore, the two densities of states at the Fermi edge E_F may differ.

of metallic and ferromagnetic properties and because the microscopic magnetic coupling mechanisms and the origin of the semimetallic character are still unclear. In this section, we try to reproduce experimentally recorded XAS and XMCD spectra [89] by means of ED simulations. This will provide new insights into the microscopic electronic properties and coupling mechanisms of the material.

CrO_2 forms the so-called rutile structure, where the chromium atoms form a bcc tetragonal lattice. In other words, there are two chromium atoms per tetragonal unit cell, which have the coordinates $(0,0,0)$ and $(\frac{a}{2},\frac{a}{2},\frac{c}{2})$. In the rutile structure each chromium atom is surrounded by an octahedron of nearest neighbor oxygen atoms, whereby the octahedra which surround neighboring chromium atoms along the c-axis share an edge. These octahedra are not ideal: while they are hardly at all elongated in z-direction, the planar square of oxygens is deformed into a rectangle.

Let us now discuss the X-ray absorbtion and X-ray magnetic circular dichroism spectra of CrO_2 in terms of the cluster model. Assuming that oxygen is in a 2^- state in this compound, chromium must be 4^+, which corresponds to a d^2 configuration. One might expect that such a strictly ionic picture is inadequate for a metallic compound like CrO_2, but it will turn out that this picture gives quite a good description of reality. The Slater integrals for the d^2 and the d^3 configuration have been computed by an atomic LDA calculation[2]. These atomic parameters are listed in Table I. Also given is the magnetic field felt by the electrons in the d-shell. It consists of the (negligible) external field plus an 'internal field' due to various types of exchange processes. From the Curie temperature of a few hundred degrees, we conclude that this field must be of the order 10^{-2} eV; it is one of the adjustable parameters in the calculation.

Another quantity which turned out to be of crucial importance for the correct description of the XMCD spectra is the crystal field acting on the chromium d-shell. The form of this crystal field in shall now be discussed in more detail. We first consider an ideal octahedron and choose a coordinate system where four oxygen atoms are at $(\pm a, \pm a, 0)$ (the so-called planar oxygens) and two oxygen atoms are at $(0,0,\pm\sqrt{2}a)$ (the two apical oxygens). In

[2]The LDA calculations were performed by Ch. Dahnken (Universität Würzburg)

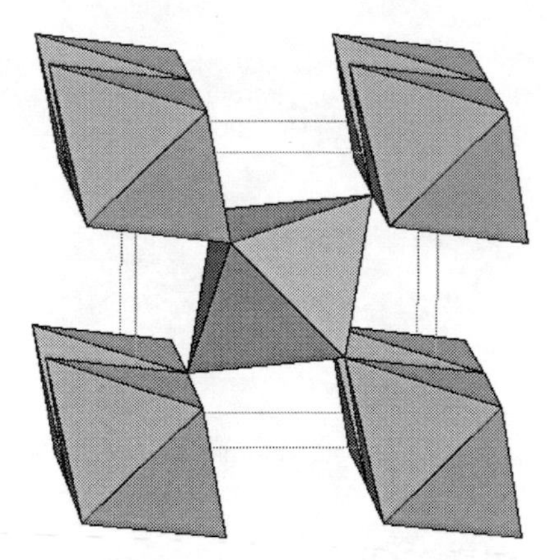

Figure 2.18: CrO_2 has a rutile crystal structure: each chromium atom is surrounded by an octahedron of nearest neighbor oxygen atoms, whereby the octahedra which surround neighboring chromium atoms along the c-axis share an edge.

a basis consisting of the 5 $l = 2$ functions $(d_{xy}, d_{yz}, d_{xz}, d_{x^2-y^2}, d_{3z^2-r^2})$ the matrix of the crystal field potential in this environment takes the well-known form

$$V_{CEF} = \begin{pmatrix} A & 0 & 0 & 0 & 0 \\ 0 & 0 & 0 & 0 & 0 \\ 0 & 0 & 0 & 0 & 0 \\ 0 & 0 & 0 & 0 & 0 \\ 0 & 0 & 0 & 0 & A \end{pmatrix}$$

This means that the $l = 2$ level is split into the three-fold degenerate t_{2g} and the two-fold degenerate e_g sub-levels. Next, let us assume that the octahedron is elongated in z-direction, i.e. the coordinates of the apical oxygens are changed to $(0, 0, \pm(\sqrt{2}a + \epsilon)$. the CEF matrix now would change to

$$V_{CEF} = \begin{pmatrix} A+b & 0 & 0 & 0 & 0 \\ 0 & 0 & 0 & 0 & 0 \\ 0 & 0 & 0 & 0 & 0 \\ 0 & 0 & 0 & 0 & 0 \\ 0 & 0 & 0 & 0 & A-b \end{pmatrix}$$

where in general $b \ll A$ (for weak distortion). In other words, the e_g sub-level is split by the lowering of the symmetry, whereas the t_{2g} level maintains its degeneracy. Precisely this type of elongation of the octahedra is realized in the CuO_2 planes of cuprate superconductors, where it leads to a unique $d_{x^2-y^2}$ ground state for the single hole in the d^9 configuration (in a coordinate system rotated by 45° with respect to the axes we use above). In the present case, however, the distance between the central chromium atom and any of its six oxygen ligands is practically the same - hence we will put the coefficient

$b = 0$.
Now we assume that the planar square of oxygen is deformed into a rectangle, in other words the coordinates of these oxygens change from $(a, a, 0)$ to $(a - \delta, a, 0)$ and accordingly for the other oxygen atoms. This changes the CEF matrix to

$$V_{CEF} = \begin{pmatrix} A & 0 & 0 & 0 & 0 \\ 0 & 0 & 0 & 0 & 0 \\ 0 & 0 & 0 & 0 & 0 \\ 0 & 0 & 0 & c & d \\ 0 & 0 & 0 & d & A \end{pmatrix}$$

If one uses δ-functions at the positions of the oxygen atoms to compute the CEF matrix, c and d turn out to be extremely small.
Now comes the crucial point of our considerations: we see that the CEF-matrix still has (for $c > 0$ and not too large d) a two-fold degenerate ground state. In fact, for small c and d the ground state is 'almost' three-fold degenerate. Putting two electrons into the d-shell (as appropriate for the Cr^{4+} ion) it turns out (by ED) that in the absence of spin-orbit coupling in the d-shell the ground state with the above CEF-matrix is a 5-fold degenerate spin triplet. This means that even the extremely weak spin-orbit interaction in the d-shell ($J_{spin-orbit} = 0.054\ eV!$) can induce a substantial orbital moment $\langle L_z \rangle$. The reason is that due to the ground state degeneracy one has to diagonalize the secular determinant of $H_{spin-orbit}$ in the subspace of the 5 degenerate ground states to obtain the ground state in the presence of $H_{spin-orbit}$. The resulting linear combination then is *independent of the magnitude* of $J_{spin-orbit}$ and hence can produce a rather large $\langle L_z \rangle$ opposite to $\langle S_z \rangle$. This will minimize the expectation value $\langle L_z S_z \rangle$. In fact, ED with a CEF matrix computed from δ-functions of strength $\approx 1 - 2\ eV$ at the position of the oxygen atoms yields a ground state expectation value of $\langle L_z \rangle \approx 0.3 - 0.5$. This is in clear contradition with experiment, where the values of $\langle L_z \rangle$ inferred from the XMCD sum-rule [90, 91] are of the order $0.02 - 0.08\hbar$. We can conclude that the simple point charge model, which would produce the above CRF matrix, is insufficient to properly describe the environment of the chromium atom. With this in mind, we increase the parameters c and d *ad hoc* to significantly larger values. In principle we might also introduce (*ad hoc*) values for the other diagonal matrix elements of V_{CEF}:

$$V_{CEF} = \begin{pmatrix} A+b & 0 & 0 & 0 & 0 \\ 0 & e & 0 & 0 & 0 \\ 0 & 0 & f & 0 & 0 \\ 0 & 0 & 0 & c & d \\ 0 & 0 & 0 & d & A-b \end{pmatrix}$$

This modification lifts the near-degeneracy of the ground state, and produces (for large d and negative c) a unique d^2 ground state which is energetically well separated from the first excited state. Consequently, the orbital moment $\langle L_z \rangle$ is substantially 'quenched' (in agreement with experiment), whereas the spin moment retains its value close to unity. The XAS and XMCD spectra computed with this modified CEF matrix then are in very good agreement with experiment, as can be seen in Figures 2.19 and 2.20, whereas they differ substantially if computed with the 'unmodified' CEF matrices computed from the point charge model.

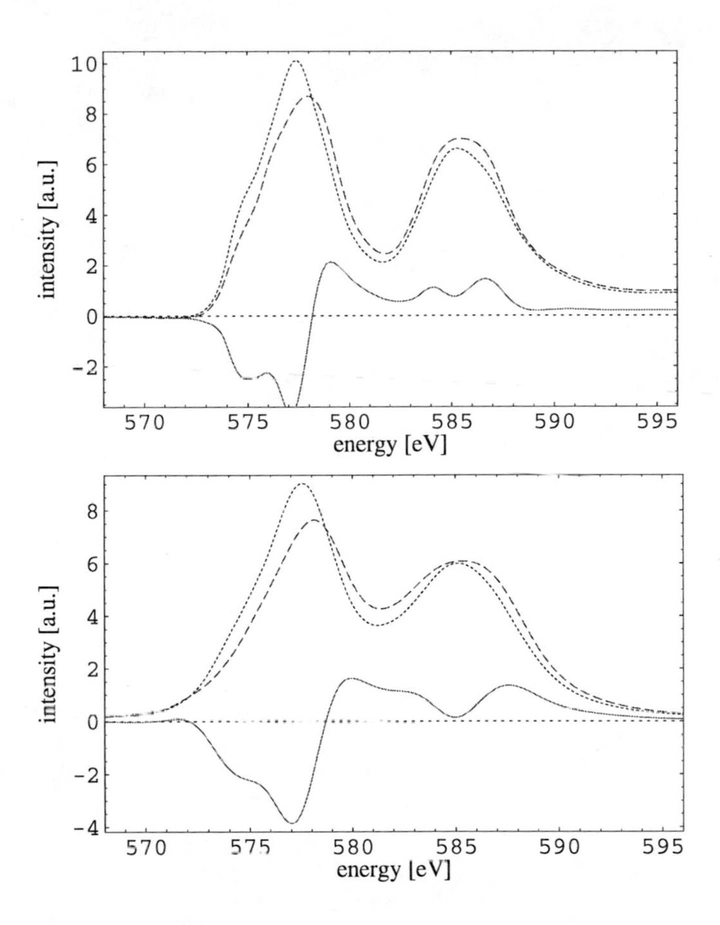

Figure 2.19: Top: experimental XAS spectra of CrO_2 obtained with left circular polarized (dashed line) resp. right circular polarized (dotted line) X-rays; the direction of the incident X-ray was parallel to the **a** *axis of the crystal lattice. The solid line represents the corresponding XMCD spectrum, i.e. the difference between the dashed and the dotted line. In the plot, the XMCD spectrum has been magnified by a factor of 2 with respect to the two XAS spectra. Bottom: the corresponding XAS and XMCD spectra obtained from an ED simulation of one $Cr0_6$ octahedron. The orbital energy difference between the chromium's 2p and 3d states has been set to 579.7 eV.*

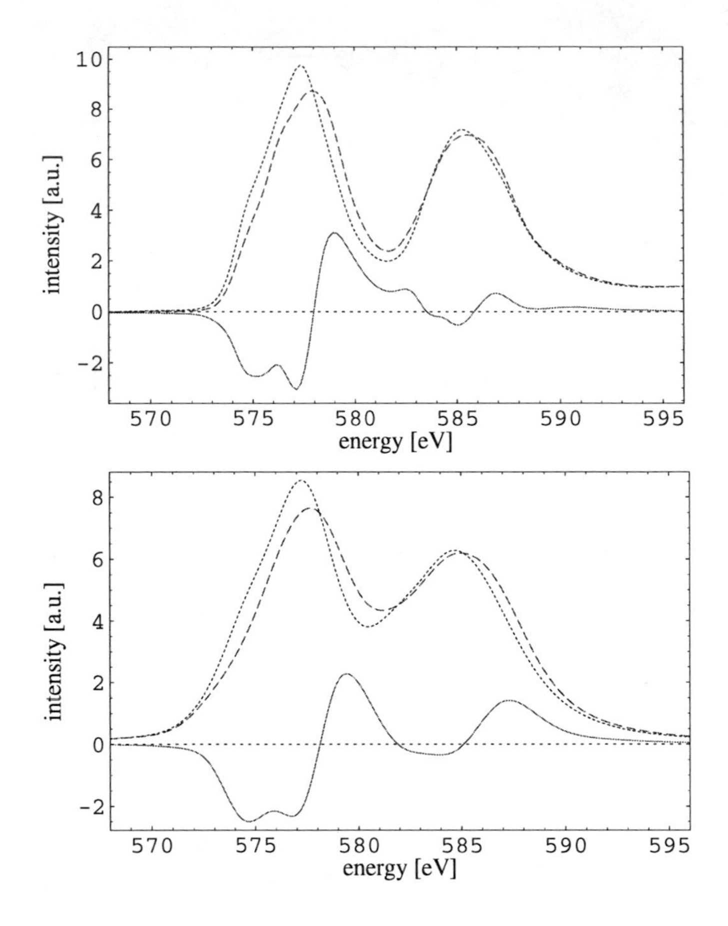

*Figure 2.20: Top: experimental XAS spectra of CrO_2 obtained with left circular polarized (dashed line) resp. right circular polarized (dotted line) X-rays; the direction of the incident X-ray formed an angle of 30° with the the **c** axis of the crystal lattice. The solid line represents the corresponding XMCD spectrum, i.e. the difference between the dashed and the dotted line. In the plot, the XMCD spectrum has been magnified by a factor of 2 with respect to the two XAS spectra. Bottom: the corresponding XAS and XMCD spectra obtained from an ED simulation of one $Cr0_6$ octahedron. The orbital energy difference between the chromium's 2p and 3d states has been set to 579.7 eV.*

In particular, not only the shape of individual spectra but also the overall trends when changing the direction of the magnetic field (which coincides with the direction of incidence of the X-rays) are reproduced reasonably well. We conclude that this modification of V_{CEF} obviously selects the correct ground state of the Cr^{4+} ion.
The additional parameters e and f might be used to improve the agreement of the calculated spectra with experiment, but it turned out that setting $e = f = 0$ gives the best agreement.

$F^0_{d,d}$	–	$F^0_{p,d}$	–	$G^1_{p,d}$	4.56
$F^2_{d,d}$	9.39	$F^2_{p,d}$	5.97	$G^3_{p,d}$	2.48
$F^4_{d,d}$	5.91	$F^4_{p,d}$	–		
[0.5mm] $J^{(2p)}_{so}$	5.67	$J^{(3d)}_{so}$	0.054	$\mu\, B_{field}$	0.03

Table 2.3: Values of Slater integrals, spin-orbit coupling constants and the magnetic field used in the atomic multiplet calculation. All values are given in eV.

In summary, we have demonstrated that the experimentally measured XAS and XMCD spectra of the magnetic compound CrO_2 can be reproduced with high accuracy by ED single particle spectra of an octahedron cluster consisting of one chromium and six oxygen atoms. This comparison of numerics and experiment permits to get a much better insight into the nature of the ground state of the system. Furthermore, it allowes to determine the values of important microscopic coupling and interaction parameters. Since our ED codes are in no way restricted on chromium dioxide, similar studies are easily feasible for many other interesting metal oxide compounds.

3

Stochastic Series Expansion (SSE)

This chapter is dedicated to the Quantum Monte Carlo simulation technique of Stochastic Series Expansion (SSE). After giving a short overview on the basic concepts of SSE, we describe a substantial new feature of the technique which has been developed during this thesis: a new method of measuring arbitrary Green's functions during the loop operator update of SSE. Then a couple of benchmark calculations is performed in which the SSE implementation written within this thesis is compared to another very performant QMC variant with nonlocal update scheme: the loop algorithm. We end this chapter by presenting a few exemplary applications which demonstrate the flexibility and wide applicability of the SSE method.

3.1 The SSE technique

3.1.1 Power series expansion of the partition function

Since their first formulation in the early eighties [52, 53], Quantum Monte Carlo (QMC) methods have become one of the most powerful numerical simulation techniques and tools in many-body physics. The first QMC algorithms were based on a discretization in imaginary time ('Trotter decomposition') and used purely local update steps to sample the system's statistically relevant states. These methods require a delicate extrapolation to zero discretization in order to reduce systematic errors. Furthermore, the purely local updates often proove incapable to traverse the accessible states in an efficient way: autocorrelation times grow rapidly with increasing system size.
A more recent class of QMC algorithms, the so-called 'loop algorithms' [19, 92–98], use non-local cluster or loop update schemes, thus reducing autocorrelation times by several orders of magnitude in some cases. Unfortunately, it is often highly non-trivial to construct a loop algorithm for a new Hamiltonian, and some important interactions cannot be incorporated into the loop scheme. These interactions have to be added as *a posteriori* acceptance probabilities after the construction of the loop, which can seriously decrease overall efficiency of the simulation. Some loop algorithms also suffer from 'freezing' [19, 99] when the probability is high that a certain type of cluster occupies almost the whole

system.
These insufficiencies can be overcome using the Stochastic Series Expansion (SSE) approach together with a loop-type updating scheme (see [16] and earlier works referenced therein).

- SSE is (almost) as efficient as loop algorithms on large systems.
- It is a numerically exact method without any discretization error.
- It is as easy to construct and general in applicability as world-line methods.

Following Sandvik [16, 17, 100], we briefly outline the basic ideas of SSE now. The central quantity to be sampled in a QMC simulation is the partition function

$$Z = \mathrm{Tr}(\mathrm{e}^{-\beta H}), \tag{3.1}$$

where H is the system's Hamiltonian and $\beta = 1/T$ the inverse temperature. Standard QMC techniques [18] split up the exponential into a product of many 'imaginary time slices' $\mathrm{e}^{-\Delta\tau H}$ and truncate the Taylor expansion of this expression after a certain order in $\Delta\tau$, thereby introducing a discretization error of order $\Delta\tau^n$. In SSE, however, one chooses a convenient Hilbert base $\{|\alpha\rangle\}$ (for example the S^z eigenbase $\{|\alpha\rangle\} = \{|S_1^z, S_2^z, ..., S_N^z\rangle\}$) and expands Z into the power series

$$Z = \sum_\alpha \sum_{n=0}^{\infty} \frac{(-\beta)^n}{n!} \langle\alpha|H^n|\alpha\rangle. \tag{3.2}$$

The statistically relevant exponents of this power series are centered around

$$\langle n\rangle \propto N_s\beta, \tag{3.3}$$

where N_s is the number of sites (or orbitals) in the system. (This follows from Eq. (3.10) and from $\langle E\rangle \propto N_s$.) We can thus truncate the infinite sum over n at a finite cut-off length $L \propto N_s\beta$ without to introduce any systematic error for practical computations. The best value for L can be determined and adjusted during an initial thermalization phase of the QMC simulation: beginning with a relatively small value of L, one can start the QMC update process, stop it whenever the cut-off L is exceeded and restart with L increased by 10...20%.
Now let H be composed of a certain number of elementary interactions involving one or two sites (such as on-site potentials, nearest neighbor hopping etc.). In order to obtain a uniform notation, we combine those interactions affecting only one site to new 'bond' interactions. One can, for example, take two chemical potential terms $\mu \cdot \hat{n}(\text{site1})$ and $\mu \cdot \hat{n}(\text{site2})$ and form the bond term $\frac{1}{C}\mu(\hat{n}(\text{site1}) + \hat{n}(\text{site2}))$ with the constant C assuring that the sum over all new bond terms equals the sum over all initial on-site terms. We can thus assume in the following that H is a finite sum of bond terms H_b and that the operator strings H^n in (3.2) can be split into terms of the form

$$\prod_{i=1}^{n} H_{b_i}^{(a_i)}, \tag{3.4}$$

where b_i labels the bond on which the elementary interaction term operates and a_i the operator type (e.g. density–density interaction or hopping). By introducing "empty" unit operators $H^{(0)} = id$, one can artificially grow all operator strings to length L and obtains [17]

$$Z = \sum_{\alpha} \sum_{\{S_L\}} \frac{\beta^n (L-n)!}{L!} \langle \alpha | \prod_{i=0}^{L} (-H_{b_i}^{(a_i)}) | \alpha \rangle. \tag{3.5}$$

Here, $\{S_L\}$ denotes the set of all concatenations of L bond operators $H_b^{(a)}$ and n is the number of non-unit operators in S_L.
If we want to sample the (α, S_L) according to their relative weights with a Monte Carlo procedure, we have to make sure that the energy of each bond operator is zero or negative since, in order to fulfill detailed balance, we choose the acceptance probability p of a bond interaction to be proportional to its negative matrix element. This requires, however, that all matrix elements are non-positive. Does a simple redefinition of the zero of energy help? For the diagonal operators, we can indeed add the same negative constant C to each of them without changing the system's properties, and thus make all matrix elements negative or zero. Unfortunately, for the non-diagonal terms an equally simple remedy does not exist. If one can show, however, that such a non-diagonal operator must appear pairwise for the matrix element to be non-zero, its energy can be multiplied by -1 without to change the physics of the system. This corresponds to a gauge transformation on all lattice sites with odd parity. On non-frustrated lattices, this trick is widely applicable, which considerably increases the set of Hamiltonians suitable for SSE. If there are valid world-line configurations carrying an odd number of non-diagonal vertices with positive energy – which is typical for Hamiltonians and lattices with frustrations – only the conventional approach of dealing with the sign problem helps [53, 101, 102]: one simulates a new system with the acceptance probabilty $p' = |p|$ and obtains the estimate of a physical quantity Q in the form

$$\langle Q \rangle = \frac{\langle Q \,\mathrm{sign}\, p \rangle}{\langle \mathrm{sign}\, p \rangle}.$$

Unfortunately, $\langle \mathrm{sign}\, p \rangle$ tends to zero exponentially with increasing N_s or β, so that the computation time needed to achieve a certain accuracy exponentially increases with $N_s \cdot \beta$ and the accessible range of system sizes and temperatures is rather limited.
When simulating charged particles in a magnetic field, a more general sign problem can arise: the vector potential A adds phase factors of the form $\exp(\mathrm{i} \int A\, dr)$ to the hopping matrix elements of all charged particles. This problem is discussed in more detail in [103].

3.1.2 Update mechanism

Having outlined the basic idea of SSE, we review the non-local updating scheme proposed by Sandvik [16]. In the following figures we illustrate the procedure by means of a simple physical model: a system of two types of hard-core bosons on a 6-site chain with periodic

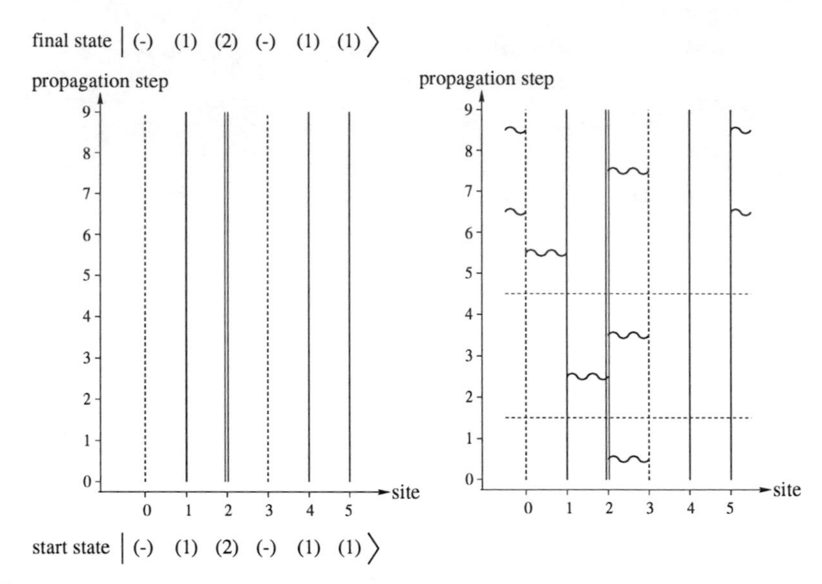

Figure 3.1: Left: world-line representation of an arbitrarily chosen start state for a physical system with three allowed occupations per site: empty (dashed line), particle 1 (solid line) or particle 2 (double line). The initial cut-off length L has been set to L = 9, and the initial bond operator string consists only of "empty" operators. Right: in the diagonal update step a certain number of empty bond operators is replaced by diagonal ones (and vice versa). In this example 7 of the initial 9 empty operators have been replaced.

boundary conditions and Hamiltonian

$$\begin{aligned} H &= -t \sum_{\alpha=1,2} \sum_i \mathcal{P} \left[a_{(\alpha)}(i)^\dagger a_{(\alpha)}(i+1) + \text{h.c.} \right] \mathcal{P} \\ &\quad + \sum_{\alpha=1,2} \mu_\alpha \sum_i n_{(\alpha)}(i) \\ &\quad + \sum_{\alpha=1,2} \eta_\alpha \sum_i \mathcal{P} \left[a_{(\alpha)}(i)^\dagger a_{(\alpha)}(i+1)^\dagger + \text{h.c.} \right] \mathcal{P}. \end{aligned} \tag{3.6}$$

The creation operators $a_{(\alpha)}(i)^\dagger$ create a hardcore boson of type $\alpha=1$ or $\alpha=2$ on site i. The first term is a nearest neighbor hopping term (t), the second term (μ_α) the chemical potential; the third term (η_α) models pair creation and annihilation. The projector $\mathcal{P} = \sum_i (1 - n_{(1)} n_{(2)})$ implements the hardcore constraints between the two types of bosons. In the world-line representation – in which the x-axis represents the spatial dimensions and the y-axis the propagation level $l = 1 \ldots L$ – we symbolize type-1 bosons by single solid lines, type-2 bosons by double lines and empty sites by dotted lines (see Fig. 3.1 left). Sandvik separates the set of all bond operators into three classes: empty operators $H^{(0)}$, diagonal operators $H^{(d)}$ and nondiagonal operators $H^{(nd)}$. The QMC process starts with an arbitrarily chosen initial state $|\,\alpha\rangle$ and an empty operator string: in Fig. 3.1 (left), for

example, three sites are occupied with type-1 bosons, two sites are empty and on site 2 is occupied by a type-2 particle. Now two different update steps are performed in alternating order: a diagonal update exchanging empty and diagonal bond operators and an operator loop update transforming and exchanging diagonal and nondiagonal operators.
In the diagonal update step, the operator string positions $l = 1...L$ are traversed in ascending order. If the current bond operator is a nondiagonal one, it is left unchanged; if it is an empty or diagonal operator, it is replaced by a diagonal or empty one with a certain probability satisfying detailed balance, i.e. an operator with lower energy is more likely to be maintained or inserted than an operator with higher energy (Fig. 3.1 right). Following Sandvik [16], we use the notation

$$|\alpha(l)\rangle = \prod_{i=1}^{l} H_{b_i}^{(a_i)} |\alpha\rangle \tag{3.7}$$

for the state obtained by acting on $|\alpha\rangle$ with the first l bond operators and $|\alpha_b(l)\rangle$ for the restriction of $|\alpha(l)\rangle$ to the bond b. Let M be the total number of interacting bonds on the lattice. Then the detailed balance conditions for the diagonal update read

$$\begin{aligned} P(H^{(0)}(l) \to H_b^{(d)}(l)) &= \min\left(1, \frac{M\beta\langle\alpha_b(l)|H_b^{(d)}|\alpha_b(l)\rangle}{L-n}\right) \\ P(H_b^{(d)}(l) \to H^{(0)}(l)) &= \min\left(1, \frac{L-n+1}{M\beta\langle\alpha_b(l)|H_b^{(d)}|\alpha_b(l)\rangle}\right). \end{aligned} \tag{3.8}$$

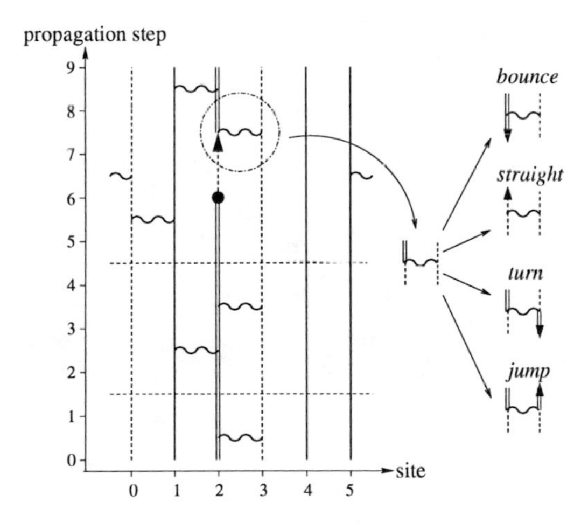

Figure 3.2: In the operator loop update step, a local change is inserted on a world-line and then moved through the world-lines and vertices. At each vertex a new direction is chosen such that the probability of a path is proportional to the negative energy of the resulting interaction vertex (detailed balance).

Nondiagonal bond operators cannot simply be inserted into the world line configuration as diagonal operators can: their insertion and modification require local changes of the world-line occupations. We discussed earlier in this paper that concatenated local changes along a closed path (or loop) through the network of world-lines and interaction vertices are much more efficient than independent purely local changes. Sandvik proposed the following method to construct such a loop (see Fig. 3.2): a certain world-line and a propagation level l on it is chosen arbitrarily; at the chosen point one disturbs the world-line by a local change – for example the creation or annihilation of a particle. Then one chooses a direction (up or down in propagation direction) and starts moving the disturbation in this direction (Fig. 3.2). The aim is to move this disturbation (we'll call it 'loop head') through the network of world-lines and interaction vertices until the initial discontinuity is reached again and healed up.

Whenever the loop head reaches an interaction vertex, we must decide how to go on; in the situation shown in Fig. 3.2, the two paths 'bounce' and 'straight' are always possible (since they result in a diagonal vertex), the path 'turn' is only allowed if the Hamiltonian contains nearest neighbor hopping terms for particle type 2, while path 'jump' is forbidden unless the Hamiltonian also allows for pair creation of particle type 2. The choice among the allowed paths must again satisfy detailed balance.

In our model – in which both pair creation and hopping are allowed – we might end up with the series 'turn', 'jump', 'turn', 'turn' of path choices, after which the starting point is regained and the world-line discontinuity healed up (Fig. 3.3). The overall result of this loop is that we have replaced 4 diagonal interactions by 4 nondiagonal interactions (marked 'n.d.') in Fig. 3.3.

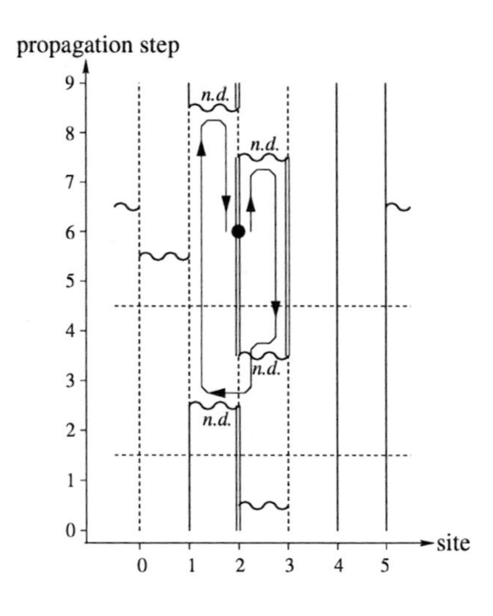

Figure 3.3: The loop update closes if the initial insertion point – here the propagation level 6 at world line 2 – is reached again and if the inserted world-line discontinuity is removed in this step.

Sandvik's method implicitly assumes that running with a world-line change into an interaction vertex always requires choosing an outgoing leg and a change on it and continuing the loop. But what if the encircled vertex in Fig. 3.2 with three empty legs and one leg occupied by particle 2 is also a valid vertex? Then we have to add a fifth possibility to the list of allowed path choices: the "don't choose an outgoing path but stop here" choice. If this last alternative is chosen, the loop has reached a dead end. In this case, our SSE code terminates the loop here, goes back to the starting point and moves in the opposite direction until either another dead end is encounterd or the starting point is reached again and the initial discontinuity is healed up.

From Fig. 3.2 one can see that when choosing a path through the current vertex, there is always the possibility to undo the current change on the incoming leg of the vertex and to 'bounce' backwards on the same leg. This path choice is normally not very helpful since it means one step backwards in the construction of the current loop. Fortunately, all 'bounce' paths can be suppressed without violating detailed balance if on each bond all nonzero matrix elements are equal, or can be made equal after a suitable energy shift of the diagonal vertices. All the Heisenberg models studied in section 3.2.1 are examples for this class of 'optimizable' physical systems.

A further improvement of the update scheme is possible in the limit of high temperatures, i.e. at $\beta \rightarrow 0$. Formula (3.3) tells us that the average number of (non-empty) vertices is rather small in this situation, and a large part of all world-lines is not connected to any vertex at all. The loop update will not be very efficient here since it essentially needs a sufficient number of vertices interconnecting the world-lines. For this reason, our SSE code additionally performs a so-called 'free world-line update' on each world line carrying no vertex at all. In this update, the occupation of the entire world-line is changed to a randomly selected new occupation.

We have stressed several times that all the local path choices satisfy detailed balance. What remains to be shown is that the whole updating mechanism is ergodic in the grand canonical ensemble, i.e. that all bond operator strings S_L and all states $|\,\alpha\,\rangle$ can be reached. In order to demonstrate that the complete update scheme is indeed ergodic, we remember that loops crossing the boundary between first and last propagation level l modify the initial state $|\,\alpha\,\rangle$ for the next update cycle. Therefore, the loops sample not only S_L but also $|\,\alpha\rangle$, and starting from a completely empty system, *any* allowed configuration can be generated by a series of loops traversing one entire world-line each.

Numerical tests of the loop-update mechanism described above show that for large values $N_s \cdot \beta$ and if there are elementary interactions with very different energy scales, the loop construction sometimes gets stuck and the loop head does not find its way back to the starting point even after millions of steps. In order to avoid this trapping, loops that exceed a critical length are aborted and the original state is restored. This causes no systematic errors for measurements done between loop updates as detailed balance is not violated. The measurements of Green's functions $G(r)$, however, which are performed 'on the run' during loop construction (see Sec. 3.1.4), are systematically biased if large loops are thrown away: since the latter are more likely to reach regions of the systems far away from the starting point than short loops, the values of $G(r)$ for large distances r are systematically under-estimated if a considerable amount of large loops is aborted. Hence, the total number of aborted loops has to be checked before one can trust in the recorded Green's functions.

3.1.3 Measurements and estimators

Efficient estimators for many static observables within the SSE mechanism have been derived by Sandvik [105].

- All observables $H^{(a)}$ appearing as elementary interactions in the Hamiltonian can be measured very easily by counting the corresponding interaction vertices in the bond operator string S_L: if S_L contains on average $\langle N(a)\rangle$ such vertices, the measured value of $H^{(a)}$ is simply

$$\langle H^{(a)}\rangle = -\frac{1}{\beta}\langle N(a)\rangle. \tag{3.9}$$

- Summing over all elementary terms $H^{(a)}$ gives an estimator for the system's internal energy:

$$E = -\frac{1}{\beta}\langle n\rangle, \tag{3.10}$$

  where n is the number of non-empty interaction vertices in S_L. (This equation can be derived very easily from $\langle E\rangle = \frac{\partial}{\partial\beta}\ln Z$.)

- For the heat capacity C_V, we additionally have to measure the fluctuations of n:

$$C_V = \langle n^2\rangle - \langle n\rangle^2 - \langle n\rangle. \tag{3.11}$$

- Equal-time correlations of two diagonal operators D_1 and D_2 can be measured via

$$\langle D_1 D_2\rangle = \left\langle \frac{1}{n}\sum_{l=0}^{n-1} d_2[l]\, d_1[l] \right\rangle, \tag{3.12}$$

  where $d_i[l]$ is the eigenvalue of D_i acting on the state $|\alpha(l)\rangle$ which results from applying the first l non-empty bond-operators $H^{(b)}$ to the initial state $|\alpha\rangle$.

Are there equally efficient estimators for time-dependent observables? In SSE the propagation index l describes the evolution of an initial state when a series of elementary terms of the Hamiltonian is acting on it; thus l plays a role analogous to imaginary time in a standard path integral. More detailed calculations [100] show that an imaginary time separation τ corresponds to a binomial distribution of propagation distances Δl; the time-dependent correlation $\langle D_2(\tau)D_1(0)\rangle$, for example, is related to the correlator

$$C_{12}(\Delta l) = \frac{1}{n}\sum_{l=0}^{n-1} d_2[l+\Delta l]\, d_1[l] \tag{3.13}$$

via

$$\langle D_2(\tau)D_1(0)\rangle = \left\langle \sum_{\Delta l=0}^{n} \binom{n}{\Delta l}\left(\frac{\tau}{\beta}\right)^{\Delta l}\left(1-\frac{\tau}{\beta}\right)^{n-\Delta l} C_{12}(\Delta l) \right\rangle. \tag{3.14}$$

The corresponding generalized susceptibilities can be calculated straight forward by numerically integrating $\langle D_2(\tau)D_1(0)\rangle$ over τ,

$$\chi_{12} = \int_0^\beta \langle D_2(\tau)D_1(0)\rangle \, d\tau. \tag{3.15}$$

Another method, which is numerically more stable, artificially adds an external field $h_2 D_2$ to the Hamiltonian and calculates the response function

$$\chi_{12} = \frac{\partial\langle D_1\rangle}{\partial h_2}\bigg|_{h_2=0}. \tag{3.16}$$

The result is [100]

$$\chi_{12} = \left\langle \frac{\beta}{n(n+1)} \left(\sum_{l=0}^{n-1} d_2[l]\right) \left(\sum_{l=0}^{n-1} d_1[l]\right) + \frac{\beta}{n(n+1)} \sum_{l=0}^{n-1} d_2[l]\, d_1[l] \right\rangle. \tag{3.17}$$

3.1.4 Measuring Green's functions

The observables listed listed in section 3.1.3 serve to access important static thermodynamic properties of the studied system. However, properties such as photo emission $\langle a^\dagger{}_{(\mathbf{k},\omega)}\, a_{(0,0)}\rangle$ or spin flip $\langle S^-{}_{(\mathbf{k},\omega)}\, S^+{}_{(0,0)}\rangle$ are often even more interesting as they provide insights into the system's dynamics. Within the framework of SSE, measuring these Green's functions $G(\mathbf{k},\omega)$ requires the insertion of local changes on certain world-lines (such as removing a particle at propagation level l_1 on world-line w_1 and re-inserting it at propagation level l_2 on world-line w_2). Performing these insertions is a highly non-trivial task since on the one hand detailed balance must be assured, on the other hand the whole process has to sample all distances $r = w_2 - w_1$ and all propagation differences $\Delta l = l_2 - l_1$ efficiently. Both requirements also exist for the QMC update steps between different measurements, and within SSE they are fulfilled by introducing the non-local operator loop update mechanism. Since this update inserts and moves local changes on the network of world-lines and connecting interaction vertices, it can be used to record the corresponding Green's functions $G(r,\Delta l)$ 'on the run' while constructing the closed loop. As an example we reconsider the hard-core boson model from section 3.1.2 and in particular the operator loop shown in Figs. 3.2 and 3.3 which starts with the removal of a type-2 particle on propagation level 6 of world-line 2; our cut-off power in the series expansion was $L = 9$, and previous diagonal updates have produced $n = 7$ 'non-empty' interaction vertices.
Taking level 6, the starting point of the loop, as zero point for the (periodically closed) propagation direction, we are now able to measure quantities of type $\langle a_1^\dagger{}_{(r,\Delta l)}\, a_2{}_{(0,0)}\rangle$ and $\langle a_2^\dagger{}_{(r,\Delta l)}\, a_2{}_{(0,0)}\rangle$ during the construction of this loop. Fig. 3.4 shows that for $\Delta l = 0$ exactly 2 measurements of $\langle a_2^\dagger{}_{(r,\Delta l)}\, a_2{}_{(0,0)}\rangle$ and one of $\langle a_1^\dagger{}_{(r,\Delta l)}\, a_2{}_{(0,0)}\rangle$ can be performed during the loop: one at the start (or end) of the loop at distance $r = 0$, two on adjacent world-lines ($r = 1$) while moving down (right) and up (left). The recorded value at each measurement is the product of the matrix elements of the first and the second change for the current world-line occupations. (For the creation or annihilation of a fermion, these matrix elements must be 1, for spin flips or creation/annihilation of bosons they can adopt other values.)

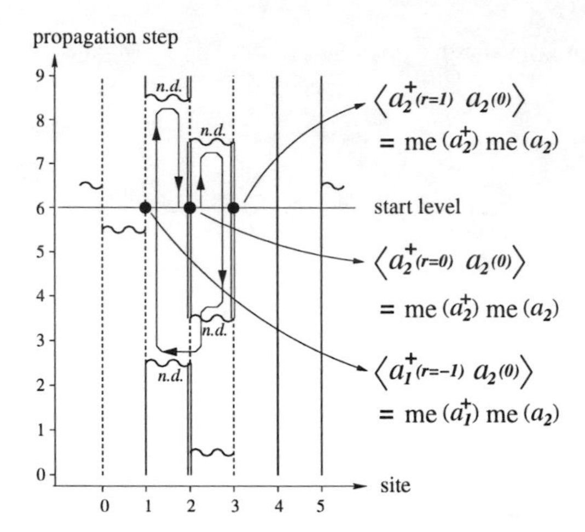

Figure 3.4: The loop update constructed in Figs. 3.2 and 3.3 can be used to record measurements of the Green's function $\langle a_2^\dagger(r,\Delta l)\, a_2(0,0)\rangle$ *and* $\langle a_1^\dagger(r,\Delta l)\, a_2(0,0)\rangle$, *where* r *is a distance between world-lines (or sites) and* Δl *is a propagation level difference. For sake of clarity only the measurements for* $\Delta l = 0$ *are explicitly marked in the figure. 'me*$(a_t^\dagger)$*' and 'me*(a_t)*' are the matrix elements for creating and annihilating a type-t particle.*

Having measured and somehow recorded the quantities $G(r, \Delta l)$ (or a correlation function $C(r, \Delta l) = \langle D_2(r, \tau) D_1(0,0)\rangle$), we still have to perform a couple of non-trivial transformation steps till we obtain the desired quantities $G(k, \omega)$ and $C(k, \omega)$ which describe the dynamical response of the system to external perturbations. First we have to relate propagation levels Δl to imaginary times τ, then a Fourier transform brings us from r-space to k-space; finally we need an inverse Laplace transform to step from imaginary time τ to exitation energy ω.

3.1.5 Efficiently accessing the system's dynamics

In this section we will discuss efficient implementation strategies for recording $G(r, \Delta l)$ and for the adjacent transformation steps mentioned above.
The transformation from propagation levels Δl to imaginary time τ requires the same weight factors as discussed earlier for diagonal correlation functions:

$$\begin{aligned} G(r,\tau) &= \sum_{\Delta l=0}^{n} \binom{n}{\Delta l} \left(\frac{\tau}{\beta}\right)^{\Delta l} \left(1-\frac{\tau}{\beta}\right)^{n-\Delta l} G(r,\Delta l) \qquad (3.18) \\ &=: \sum_{\Delta l=0}^{n} w(\tau,\Delta l)\, G(r,\Delta l)\,. \end{aligned}$$

Since in SSE the artificially introduced empty vertices are statistically uniformly distributed amidst the non-empty vertices, Eq. (3.18) remains valid if one replaces n, the

fluctuating number of non-empty vertices, by L, the fixed total number of vertices. For practical computations, this second form of Eq. (3.18) is more convenient because the binomial weight prefactors are now fixed during the entire simulation and can easily be calculated once at the beginning of the simulation.
There are several possible ways to implement the recording of $G(r, \Delta l)$ measurements and the adjacent transformation to $G(r, \tau)$. The easiest and at first glance fastest way simply writes all recorded $G(r, \Delta l)$ data into a two dimensional array with dimensions N_s and $L \propto N_s\beta$. The transformation to $G(r, \tau)$ can then be performed once at the end of the simulation. This method raises two problems. First, in principle a separate measurement has to be recorded each time the loop head steps up or down by one level on a world-line and whenever it traverses an interaction vertex. Recording all these single measurements drastically slows down the loop update process. Second, for large systems (e.g. $N_s \approx 5000$) and low temperatures (e.g. $\beta \approx 40$), the two dimensional array needed to store $G(r, \Delta l)$ contains about 1 billion elements and needs approximately 10 GB of memory – more than available on many computer systems, and certainly more than the fast cache memory can store.
In order to overcome these problems, one can replace the 'brute force' recording of data on *all* traversed $(r, \Delta l)$ points by a Monte Carlo sampling: in each loop-update a distance Δl is chosen randomly, according to the probabilities in Eq. (3.18), for each of the times τ of interest. Measurements are then performed only at these Δl and transformed directly into τ.
In our code we have adopted a third strategy: we perform *all* possible $G(r, \Delta l)$ measurements, thereby exploiting the fact that $G(r, \Delta l)$ is constant on the entire world-line fragment between tho adjacent vertices, and directly transform these 'raw data' into $G(r, \tau)$ at the end of each loop update step. The transformation after each QMC update step is necessary to keep memory requirements low. Simply applying Eq. (3.18) with its 'expensive' operations (divisions, powers, binomial coefficients, large sums) would cost by far too much computation time. Instead, we remember that $G(r, \Delta l)$ is composed of a relatively small number of intervals $I =]\Delta l_1, \Delta l_2]$ with constant function value (Fig. 3.5b)). Therefore, we can compute the contribution of an entire Δl-interval to $G(r, \tau)$ in one step:

$$G(r, \tau) = \sum_I G(r, I)\big(W(\tau, \Delta l_{2(I)}) - W(\tau, \Delta l_{1(I)})\big), \tag{3.19}$$

where W is the 'integrated weight function'

$$W(\tau, \Delta l) = \sum_{m=0}^{\Delta l} w(\tau, m). \tag{3.20}$$

The Δl-range in which $W(\tau, \Delta l)$ considerably differs from 0 and 1, is determined by mean value and standard deviation of the binomial distribution $w(\tau, \Delta l)$

$$\langle \Delta l \rangle = L\frac{\tau}{\beta} \tag{3.21}$$

$$\sigma_{\Delta l} = \sqrt{L\frac{\tau}{\beta}\left(1 - \frac{\tau}{\beta}\right)}. \tag{3.22}$$

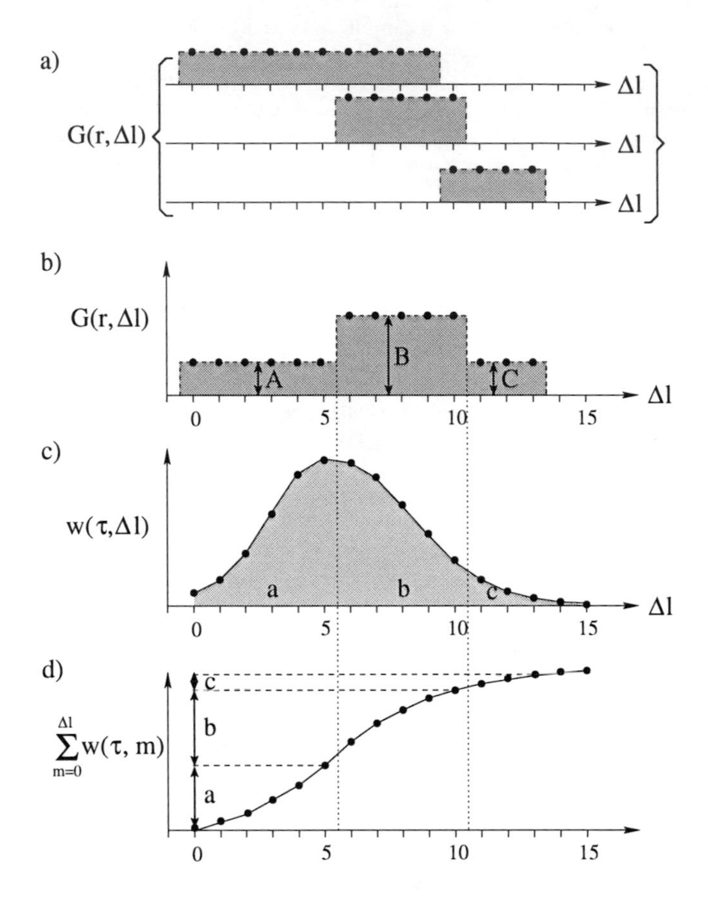

Figure 3.5: Transformation of Green's functions measurements from propagation level Δl to imaginary time τ: the raw measurements recorded during loop update on different world-line segments (a) are combined into a single function $G(r, \Delta l)$ (b). For a given τ, $G(r, \tau)$ could be computed by summing up all $G(r, \Delta l)$ weighted with $w(\tau, \Delta l) = \binom{L}{\Delta l}\left(\frac{\tau}{\beta}\right)^{\Delta l}\left(1-\frac{\tau}{\beta}\right)^{L-\Delta l}$ (c). A much more efficient way uses the 'integrated weight function' $W(\tau, \Delta l) = \sum_{m=0}^{\Delta l} w(\tau, m)$ (d) to get the total contribution of each range $]\Delta l_1, \Delta l_2]$ in which $G(r, \Delta l)$ is constant. In the example shown here $G(r, \tau)$ is then simply $aA + bB + cC$.

Below $\langle \Delta l \rangle - 5\,\sigma_{\Delta l}$ the integrated weight is zero, above $\langle \Delta l \rangle + 5\,\sigma_{\Delta l}$ it is 1 (up to an error of less than 10^{-7}). The remaining interval rarely contains more than fifty or hundred Δl-points (see Fig. 3.5d); these values can easily be stored after having been computed once for each τ. Thus, $W(\tau, \Delta l)$ can be calculated very rapidly with nothing but a couple of 'cheap' elementary operations.
For very large system sizes of 10000 or more sites and very low temperatures, the 'relevant' Δl-ranges might become so large that it is unfavorable to store all needed $W(\tau, \Delta l)$ values – for example because accessing the large array $W[\tau_i, \Delta l]$ would caust too many cache misses. In this case, one can store the coefficients of some interpolation functions for $W(\tau, \Delta l)$ instead of the function values themselves. Practical tests have shown that dividing the relevant interval $[\langle \Delta l \rangle - 5\sigma_{\Delta l}, \langle \Delta l \rangle + 5\sigma_{\Delta l}]$ into six sub-intervals with boundaries $\langle \Delta l \rangle - 5\,\sigma_{\Delta l}$, $\langle \Delta l \rangle - 2.8\,\sigma_{\Delta l}$, $\langle \Delta l \rangle - 1.3\,\sigma_{\Delta l}$, $\langle \Delta l \rangle$, $\langle \Delta l \rangle + 1.3\,\sigma_{\Delta l}$, $\langle \Delta l \rangle + 2.8\,\sigma_{\Delta l}$ and $\langle \Delta l \rangle + 5\,\sigma_{\Delta l}$ and interpolating W in each sub-interval by a fifth-order polynomial is a good compromise between evaluation speed (about 15 elementary operations), storage requirements (36 floating point numbers for each τ) and interpolation accuracy (better than $2..3 \times 10^{-7}$).
The next transformation step, Fourier transform from $G(r)$ to $G(k)$, is a well known standard method that does not impose any fundamental problems. However, standard Fast Fourier Transform (FFT) algorithms perform best if *all* $G(k)$ values are to be calculated, whereas in practice one rarely needs all k-values and is interested only in one k-point or in some special points of the Brillouin zone, e.g. the point $k = (\pi, \pi)$ and its immediate neighborhood. In this situation, one can save a lot of computation time by not recurring to FFT but using optimized algorithms designed particularly for these 'selective' Fourier problems. If we are interested in only one or a few k-points, we can use a simple Fourier transform to get $\{G(k, \tau)\}$ from $\{G(r, \tau)\}$ in $\mathcal{O}(N_s \cdot n_k)$ operations, with n_k being the number of k-points. Correlation functions $C(k, \tau)$ can even be measured directly in k-space, which also can be done in $\mathcal{O}(N_s{\cdot}n_k)$ operations. For the case $1 \ll n_k \ll N_s$ we have implemented a new FT method performing much better than FFT in this situation [106].
Unlike a Fourier transform, a Laplace transform in general cannot be inverted. Therefore, the last transition step from τ to ω is by far more complicated than the previous one from r to k. We use *Maximum Entropy* techniques developed within the last years and refer to earlier publications [107].

3.2 Scaling behavior and benchmark tests

3.2.1 Scaling behavior

One decisive criterium for the performance of a QMC simulation technique is the behavior of computation time C as a function of system size N_s or inverse temperature β. To facilitate a hardware-independent measurement of C and a comparison to other QMC techniques, we define C as the number of elementary update operations needed to transform a given state $|\alpha^{(n)}\rangle$ into an new state $|\alpha^{(n+1)}\rangle$ in such a way that the mean autocorrelation time τ is equal to 1. In SSE the number of elementary update operations is the number of diagonal vertices tested for replacement plus the number of vertices traversed during the loop update. In the following, we compare SSE to the loop algorithm, which is known to show an excellent scaling behavior for many benchmark problems. As test models we choose isotropic antiferromagnetic Heisenberg models in 1, 2 and 3 dimensions with up to 4096 sites and β up to 64 in a vanishing or finite external magnetic field. Following [104], we describe the scaling behavior of the two algorithms by means of the dynamical exponent z defined from

$$\tau{\cdot}C \propto \beta{\cdot}l^D l^z \,. \tag{3.23}$$

Here, $\tau{\cdot}C$ is the computational effort (i.e. the number of elementary update steps) needed to achieve a mean autocorrelation time of $\tau = 1$ for the measurements of the studied quantity; D is the spatial dimension of the simulated system and $l = \sqrt[D]{N_s}$ its length in each dimension. From Table 3.1 we see that both simulation techniques show an approximately equal performance and a very good scaling behavior: since the ratio $C\tau/(\beta N_s)$ is approximately constant, we obtain $z \approx 0$ in both cases.

$\beta \cdot N_s$	SSE		Loop	
	$\frac{C \cdot \tau}{\beta \cdot N_s}$	χ	$\frac{C \cdot \tau}{\beta \cdot N_s}$	χ
$4 \cdot 4^2$	1.00	$0.040(46 \pm 16)$	1.00	$0.040(20 \pm 10)$
$8 \cdot 8^2$	0.61	$0.044(83 \pm 15)$	0.90	$0.044(92 \pm 8)$
$16 \cdot 16^2$	0.40	$0.044(72 \pm 12)$	0.56	$0.044(68 \pm 6)$
$32 \cdot 32^2$	0.40	$0.044(19 \pm 11)$	0.53	$0.044(24 \pm 6)$
$64 \cdot 64^2$	0.36	$0.044(01 \pm 23)$	0.42	$0.044(07 \pm 14)$

Table 3.1: 2D antiferromagnetic Heisenberg model at vanishing magnetic field $h = 0$: calculation of the uniform magnetic susceptibility $\langle\chi\rangle = \frac{\partial\langle M\rangle}{\partial h}\big|_{h=0}$ from QMC simulations with 1000000 (120000 in the case $\beta = L = 64$) update-measurement cycles (left: SSE, right: loop algorithm). $C \cdot \tau$ is the number of elementary update operations per cycle needed to achieve a mean autocorrelation time $\tau = 1$ for the measurements of χ.

Next, we enlarge the square lattice into the third spatial dimension and examine a bilayer with open boundary conditions in z-direction and periodic boundary conditions in (xy). From QMC simulations with the loop algorithm we know that the quantum critical point – separating a spin gap phase at large coupling $J_\perp$ in z-direction from a phase with long-range order at small $J_\perp$ – is situated at $J_\perp/J = 2.524$. Our aim is to measure scaling behavior and dynamical exponents exactly at this quantum critical point. This point is of particular interest since the immediate neighborhood of a phase transition often leads to the so-called 'critical slowing down' of QMC simulations, i.e. exploding autocorrelation times and thus a dramatic decrease of efficiency of the QMC update process.
The results in Table 3.2 show that the scaling behavior for both algorithms is still almost linear in βN_s. The scaling for SSE looks slightly superior to the loop algorithm. This difference can most probably be attributed to the fact that improved estimators were used in the loop algorithm simulation, leading to slightly smaller errors but larger autocorrelation times. There is no sign of critical slowing down in either algorithm.
As we have mentioned in the introduction of this paper, one of the major advantages of SSE is that external potentials and fields can be included without a loss of performance.

$\beta \cdot N_s$	SSE $\frac{C \cdot \tau}{\beta \cdot N_s}$	SSE χ	Loop $\frac{C \cdot \tau}{\beta \cdot N_s}$	Loop χ
$4 \cdot 2 \cdot 4^2$	1.00	$0.0115(6 \pm 7)$	1.00	$0.0114(6 \pm 5)$
$8 \cdot 2 \cdot 8^2$	0.96	$0.0068(2 \pm 6)$	1.03	$0.0069(2 \pm 2)$
$16 \cdot 2 \cdot 16^2$	0.68	$0.0036(8 \pm 4)$	1.20	$0.0036(6 \pm 2)$
$32 \cdot 2 \cdot 32^2$	0.56	$0.0018(5 \pm 3)$	1.20	$0.0018(3 \pm 1)$

Table 3.2: Square bilayer antiferromagnetic Heisenberg model at vanishing magnetic field $h = 0$ and at the quantum critical point ($J_\perp/J = 2.524$): calculation of the uniform magnetic susceptibility $\langle\chi\rangle$ from QMC simulations with 1000000 (390000 in the case $\beta = L = 32$) update-measurement cycles (left: SSE, right: loop algorithm).

h/J	$\beta \cdot h$	SSE $\frac{C \cdot \tau}{C_0 \cdot \tau_0}$	SSE M	Loop $\frac{C \cdot \tau}{C_0 \cdot \tau_0}$	Loop M
0.02	0.32	1.00	$0.008(2 \pm 5)$	1.00	$0.0083(7 \pm 7)$
0.04	0.64	1.02	$0.016(4 \pm 6)$	0.97	$0.017(6 \pm 2)$
0.1	1.6	1.22	$0.057(8 \pm 8)$	1.84	$0.057(7 \pm 5)$
0.2	3.2	1.98	$0.24(4 \pm 2)$	7.55	$0.24(3 \pm 2)$
0.4	6.4	1.37	$0.893(8 \pm 9)$	167.40	$0.89(4 \pm 3)$
1.0	16.0	0.86	$2.069(0 \pm 8)$	2338.74	$2.(24 \pm 12)$

Table 3.3: 1D antiferromagnetic Heisenberg model in a magnetic field h: calculation of magnetization M from QMC simulations with 1000000 update-measurement cycles (left: SSE, right: loop algorithm) for a system with $\beta = N_s = 16$. $C \cdot \tau$ is the number of elementary update operations per cycle needed to achieve a mean autocorrelation time $\tau = 1$ in measuring M.

To verify this assertion, we now examine the antiferromagnetic Heisenberg model on a 1D chain and a 2D square lattice in a finite magnetic field $h > 0$. For the loop algorithm we expect to find a rapidly increasing autocorrelation time and decreasing performance if the product of magnetic field h and inverse temperature is much larger than 1. This is due to the fact that the external field is incorporated into the loop algorithm via a-posteriori acceptance probabilities for each constructed loop. for $\beta h \ll 1$ these probabilities are still almost equal to 1, whereas at $\beta h \approx 1$ they begin to decrease considerably.
Indeed, the numerical results from Table 3.3 demonstrate that at $\beta h \approx 2$ the autocorrelation time begins to grow rapidly, and at $\beta h \approx 10$ the algorithm cannot be used any more because the autocorrelation times get too long. For SSE, on the contrary, we do not expect any negative effect by introducing a magnetic field whose strength is of the order of the other elementary interactions, $h \approx J$, since no a-posteriori acceptance decision is neccessary. We rather presume that performance is slightly worse for $h/J \ll 1$ because there are elementary interaction vertices with very different energy scales. Both predictions are verified by the data in Table 3.3.
For sake of completeness we also show the corresponding data for the 2-dimensional

h/J	$\beta \cdot h$	SSE $\frac{C \cdot \tau}{C_0 \cdot \tau_0}$	SSE M	Loop $\frac{C \cdot \tau}{C_0 \cdot \tau_0}$	Loop M
0.02	0.32	1.00	$0.22(4 \pm 8)$	1.00	$0.231(3 \pm 3)$
0.04	0.64	0.83	$0.47(9 \pm 8)$	1.38	$0.477(7 \pm 7)$
0.1	1.6	0.94	$1.41(0 \pm 9)$	5.61	$1.42(2 \pm 2)$
0.2	3.2	0.44	$3.47(0 \pm 7)$	34.25	$3.48(6 \pm 7)$
0.4	6.4	0.18	$7.73(7 \pm 4)$	1691.66	$7.7(8 \pm 7)$
1.0	16.0	0.12	$22.10(6 \pm 4)$	– – –	– – – – – –

Table 3.4: 2D square antiferromagnetic Heisenberg model in a magnetic field h: calculation of magnetization M from QMC simulations with 1000000 update-measurement cycles (left: SSE, right: loop algorithm) for a system with $\beta = 16$ and $N_s = 16^2$. $C \cdot \tau$ is the number of elementary update operations per cycle needed to achieve a mean autocorrelation time $\tau = 1$ for the measurements of M.

Heisenberg model in Table 3.4. The results demonstrate that the different behavior of SSE and loop algorithm described in the one dimensional case is even more severe in two dimensions: the loop algorithm looses its efficiency already at $\beta h \approx 1.5$.
The numerical comparisons discussed above are suitable for comparing different numerical algorithms in a hardware-independent way. For practical purposes, other performance criteria of a simulation technique are more interesting. For example, one could ask how the computation time till a certain accuracy in a certain measured variable is reached, scales with β and N_s. This type of questions is studied in Fig. 3.6. For our standard 2D AF Heisenberg model we trace the computation time to reach an accuracy of 4 digits in energy as a function of β (Fig. 3.6 top) and N_s (Fig. 3.6 bottom). The exponents $\kappa(\beta)$ in $C \propto \beta^{\kappa(\beta)}$ and $\kappa(N_s)$ in $C \propto \beta^{\kappa(N_s)}$ derived from Fig. 3.6 are

$$\kappa(\beta) = 0.34 \pm 0.05 \quad \text{and} \quad \kappa(N_s) = 0.48 \pm 0.05.$$

Both quantities are smaller than 1, and Eq. (3.23) would return a negative dynamical exponents z. This is due to self-averaging: in a large system local fluctuations of a physical observable around its mean value on different subregions of the lattice can compensate and average out each other, thereby lowering the observable's measured variance. The computational effort needed to get thermodynamical averages to a certain relative error, scales sublinearly with system size and inverse temperature, so that systems of several thousand sites or at temperatures of not more than $0.001J$ can be simulated within minutes or a few hours on a standard PC or workstation.

3.2.2 Efficacity of the Green function measurements

In this section we use our standard benchmark model – the 2D AF Heisenberg model – to test our method of Green's functions measurements for correctness and numerical

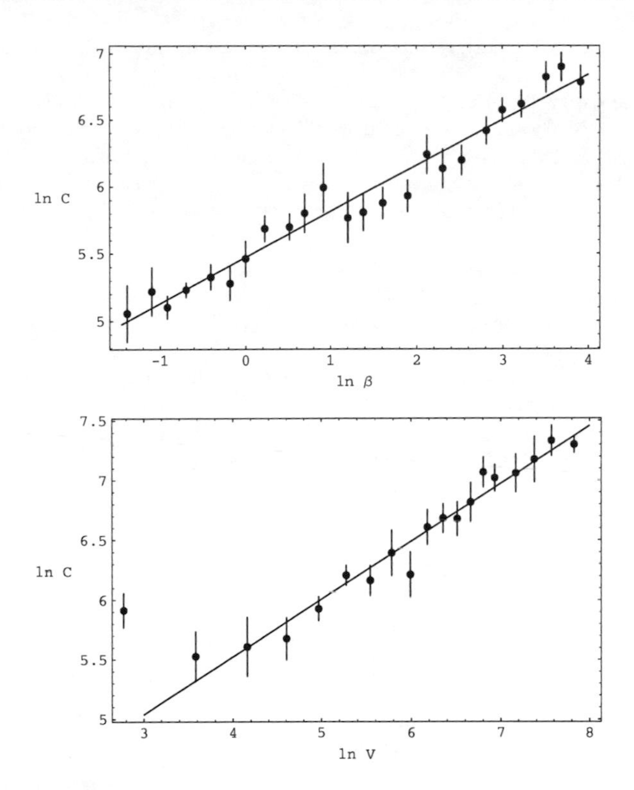

Figure 3.6: Scaling behavior of computation time C to reach a relative accuracy of 10^{-4} in the measured energy of the 2D AF Heisenberg model. On top: $\ln(C)$ versus $\ln(\beta)$ for 10×10 sites. Below: $\ln(C)$ versus $\ln(N_s)$ for $\beta = 10$. The time was measured in seconds on a DEC workstation.

efficiency. To this purpose, we calculate and compare the correlation functions

$$\langle S^z(\tau, k)\, S^z(0, k)\rangle \quad \text{and} \tag{3.25}$$

$$\langle S^+(\tau, k)\, S^-(0, k)\rangle. \tag{3.26}$$

In the S^z-eigenbase, which is normally used to span the model's Hilbert space, expression (3.25) is a time-dependent correlation function of two diagonal operators. Therefore (3.25) can be measured using estimator (3.14) without introducing changes in the world-lines and vertices defining the current state of the system. (3.26), however, consists of two nondiagonal operators and can only be measured with our new method of recording general Green's functions that was described in section 3.1.4. Furthermore, at zero field $h = 0$ both correlation functions are related via

$$\langle S^z(\tau, k)\, S^z(0, k)\rangle = \frac{1}{2}\langle S^+(\tau, k)\, S^-(0, k)\rangle, \tag{3.27}$$

so that the correctness of both estimators can be checked by directly comparing these two quantities.
When working with SSE, however, we are faced with the problem that the AF Heisenberg model (i.e. $J > 0$) possesses two nondiagonal interaction vertices with positive energy, namely $J\,S^+S^-$ and $J\,S^-S^+$. Therefore, a direct simulation of this model with SSE is impossible. Yet, observing that these two vertices must appear pairwise in any non-vanishing matrix element, we can recur to the trick mentioned at the end of section 3.1.1 and replace $S^+(\mathbf{r})$ and $S^-(\mathbf{r})$ by $(-1)^{|\mathbf{r}|}\hat{S}^+(\mathbf{r})$ and $(-1)^{|\mathbf{r}|}\hat{S}^-(\mathbf{r})$ without modifying the physical properties of the system. Here, $|\mathbf{r}|$ is defined to be even on one sublaticce and odd on the other. Assuming that there are only nearest neighbor interactions, this gauge transformation indeed solves our problem since each vertex $J\,S^+(r)S^-(r\pm 1)$ gets an additional total prefactor of -1. However, when transforming from $\langle \hat{S}^+(\tau,r)\,\hat{S}^-(0,0)\rangle$ to $\langle \hat{S}^+(\tau,k)\,\hat{S}^-(0,k)\rangle$, we have to keep in mind that the alternating gauge transformation introduces a momentum shift of $Q = (\pi,\pi)$, and (3.27) now reads

$$\langle S^z(\tau,k)\,S^z(0,k)\rangle = \frac{1}{2}\langle \hat{S}^+(\tau,k+Q)\,\hat{S}^-(0,k+Q)\rangle. \tag{3.28}$$

The numerical data in Table 3.5 perfectly fulfill this equality and hence demonstrate the correctness of our Green's functions measurements.

k	τ	$\langle S_{z(k,\tau)}\,S_{z(k,0)}\rangle$	$\frac{1}{2}\langle S^{+}{}_{(k+Q,\tau)}\,S^{-}{}_{(k+Q,0)}\rangle$
$(\frac{\pi}{2},\frac{\pi}{2})$	0	$0.16(89\pm 20)$	$0.168(2\pm 4)$
	0.1	$0.00(01\pm 17)$	$0.003(2\pm 3)$
	0.5	$-0.00(36\pm 15)$	$-0.000(4\pm 3)$
$(\frac{3\pi}{4},\frac{3\pi}{4})$	0	$0.38(39\pm 21)$	$0.386(0\pm 6)$
	0.1	$0.01(74\pm 20)$	$0.020(9\pm 5)$
	0.5	$0.00(41\pm 19)$	$0.000(4\pm 4)$
(π,π)	0	$11.3(35\pm 17)$	$11.36(1\pm 9)$
	0.1	$10.4(04\pm 17)$	$10.42(3\pm 9)$
	0.5	$9.0(83\pm 17)$	$9.09(3\pm 9)$
$(\frac{3\pi}{4},\pi)$	0	$0.53(85\pm 23)$	$0.542(2\pm 7)$
	0.1	$0.06(31\pm 20)$	$0.063(0\pm 6)$
	0.5	$0.00(38\pm 17)$	$0.000(0\pm 5)$
$(\frac{\pi}{2},\pi)$	0	$0.28(92\pm 23)$	$0.287(6\pm 5)$
	0.1	$0.01(08\pm 19)$	$0.008(5\pm 4)$
	0.5	$0.00(06\pm 17)$	$0.000(3\pm 4)$

Table 3.5: comparison of $\frac{1}{2}\langle \hat{S}^+(\tau,k+Q)\,\hat{S}^-(0,k+Q)\rangle$ and $\langle S^z(\tau,k)\,S^z(0,k)\rangle$ for the 16×16-site 2D AF Heisenberg model at $\beta = 16$ and zero magnetic field. The table shows some k-values around (π,π).

In the simulation recorded in Fig. 3.5, we have calculated $\langle S^z\,S^z\rangle$ and $\langle S^+\,S^-\rangle$ for all allowed k-points on the path $(0,0)\to(\pi,0)\to(\pi,\pi)\to(0,0)$. Table 3.5 shows a subset of these points in the vicinity of (π, π). The three tasks "performing updates", "measuring $\langle S^z S^z \rangle$" and "measuring $\langle S^+ S^+ \rangle$" contributed the following percentages to overall computation time:

performing updates	:	18.8 %
measuring $\langle S^z S^z \rangle$	:	36.1 %
measuring $\langle S^+ S^- \rangle$	:	45.1 %

From this list and the measurement accuracies in Table 3.5 we conclude that the highly non-trivial Green's functions measurements lead to a slightly better accuracy than the direct $\langle S^z(\mathbf{r}, \tau) S^z(\mathbf{r}', \tau') \rangle$ measurements while consuming roughly the same amount of computer time as the latter. Measuring the Green's function is thus the preferred method of determining also the diagonal real space dynamical correlation functions.

3.2.3 Summary

Stochastic Series Expansion (SSE) together with the implementation tricks and Green's functions measurements described in this paper is a highly performant quantum Monte Carlo simulation technique allowing to access both static and dynamical properties of very large systems of thousands of sites and at very low temperatures. Compared to the loop algorithm, which is slightly faster on big systems for some specific Hamiltonians, SSE has the advantages of not suffering from exponential slowing down in external fields; furthermore, SSE is more easily applicable to wide classes of Hamiltonians.

3.3 Exemplary applications

The SSE program codes written within this thesis have been used by several research groups to study a broad range of many-particle systems [21, 22, 27, 29, 108]. The following subsections give a short overview on these works. Additionally, SSE studies from the author's own main physical research area will be presented in chapter 6.

3.3.1 Destruction of superfluid and long-range order by impurities in two dimensional systems

K. Bernardet and G.G. Batrouni[1] have used the SSE method in collaboration with M. Troyer[2] and the author to examine the effect of impurities, in the form of disordered chemical potential, on the phase diagram of the hardcore bosonic Hubbard model in two dimensions [21]. This model is often used to study the properties of several physical systems such as Helium adsorbed on surfaces [109] and granular superconductors. The study shows that in two dimensions, no matter how weak the disorder is, it will always destroy the long range density wave order (checkerboard solid) present at half filling and strong near neighbor (nn) repulsion. In addition part of the superfluid phase surrounding the checkerboard solid is also destroyed. Properties of the glassy phase thus generated at

[1] Institut Nonlinéaire de Nice, Université de Nice
[2] Eidgenössische Technische Hochschule Zürich

strong nearest neighbor coupling, and the possibility of other localized phases at weak nn repulsion, i.e. Anderson localization, have been studied. The SSE algorithm was used to measure several physical quantities such as the superfluid density, energy gaps, and equal and unequal time Green functions.

3.3.2 Quantum phase transitions in the two dimensional hard-core boson model

F. Hebert and G.G. Batrouni, R.T. Scalettar[1], G. Schmid and M. Troyer[2], and the author have used the SSE method to map out the phase diagram of the two-dimensional hardcore boson Hubbard model with nearest (V_1) and next nearest (V_2) neighbor repulsion [22]. This Hamiltonian can be considered a model of the superconductor–insulator transition in materials with preformed Cooper pairs, [110–115] of Helium in disordered and restricted geometries, [116–118] of spin–flop transitions in quantum spin systems in external magnetic fields, [119] and of supersolid behavior. [120, 121] At half-filling three phases are detected: superfluid (SF), checkerboard solid and striped solid depending on the relative values of V_1, V_2 and the kinetic energy. Doping away from half filling, the checkerboard solid undergoes phase separation: The superfluid and solid phases co-exist but not as a single thermodynamic phase. As a function of doping, the transition from the checkerboard solid is therefore first order. In contrast, doping the striped solid away from half filling instead produces a striped supersolid phase: Co-existence of density order with superfluidity as a single phase. One surprising result is that the entire line of transitions between the SF and checkerboard solid phases at half filling appears to exhibit dynamical $O(3)$ symmetry restoration. The transitions appear to be in the same universality class as the special Heisenberg point even though this symmetry is explicitly broken by the V_2 interaction.

3.3.3 A bosonic picture for the Heisenberg spin ladder with impurites

M. Jöstingmeier, R. Eder, W. Hanke[3] and the author have studied a bosonic representation of the AF spin-1/2 Heisenberg ladder with nonmagnetic impurities (empty sites). This Hamiltonian can serve as a model for nonmagnetic Zn impurities in cuprate superconductors with depleted CuO_2 planes, which results in ladder-like arrangements of the copper atoms [122]. The electron configuration of Cu^{2+} in the CuO_2 planes of the cuprates is $[Ar]3d^9$, hence there is one hole in the $3d$ shell and Cu^{2+} has spin $S = 1/2$. Zinc has one more electron, so that Zn^{2+} has a completely filled $3d$ shell and spin $S = 0$. Zn doping in CuO_2 planes has two interesting effects. First, T_c is depressed, and the superconducting phase vanishes already at modest Zn doping concentrations of about 3% (see Fig. 3.7). Second, Zn doping induces antiferromagnetic order. A similar effect is observed in dimerized spin chains in which spinless impurities are inserted. These facts

[1]University of California, Davis
[2]Eidgenössische Technische Hochschule Zürich
[3]Universität Würzburg

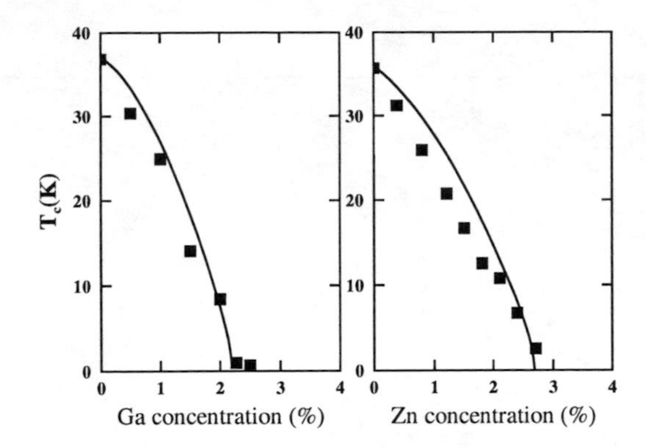

Figure 3.7: Suppression of superconductivity by zinc or calcium doping in the HTSC compound $La_{1.85}Sr_{0.15}Cu_{1-x}Ni_x0_4$. (Figure by Y.J. Kim and K.J. Chang, cond-mat/9801071)

suggest that unpaired spins induce antiferromagnetic order in systems in which the majority spins tend to form singlet pairs. Indeed, Iino and Imada [123] have shown in a QMC study of the Heisenberg ladder with spinless impurity sites that AF ordering is increased. Furthermore, they have detected a strong impact of the impurity distribution – randomly or regularly – onto the physical properties of the system.
Here, we study the bond boson representation of the Heisenberg ladder [44, 45] given in (1.12); the effect of the impurities is to create triplet excitations which are radiated off into the rest of the system [124]. More precisely, the impurities can be incorporated into the Hamiltonian by *Kondo-type couplings* between the bond bosons and the dangling copper spins on ladder rungs containing one Zn atom [124, 125]:

$$
\begin{aligned}
H_1 &= const \sum_{\langle i,j \rangle} \mathbf{S}_i \cdot (\mathbf{t}_j + \mathbf{t}_j^\dagger) \\
H_2 &= const \sum_{\langle i,j \rangle} \mathrm{i}\, \mathbf{S}_i \cdot (\mathbf{t}_j^\dagger \times \mathbf{t}_j)
\end{aligned}
\tag{3.29}
$$

For the Hamiltonian consisting of the parts (1.12) and (3.29), we study the uniform susceptibility

$$
\chi = \frac{1}{TN_s} \sum_{i,j} \langle S_i^z S_j^z \rangle \,, \tag{3.30}
$$

where N_s is the number of sites. By means of SSE we perform QMC simulations of the 64-rung ladder (which becomes a 64-site chain in the bosonic representation) at temperatures ranging down to $T = J/120$ and for 0, 2, 4, 8, and 16 impurities in the system (corresponding to Zn concentrations of 0, 1.5625%, 3.125%, 6.25%, and 12.5%). Without impurities, theory predicts [126]

$$
\chi(T) = C_0 \frac{\mathrm{e}^{-\Delta/T}}{\sqrt{T}} \,, \tag{3.31}
$$

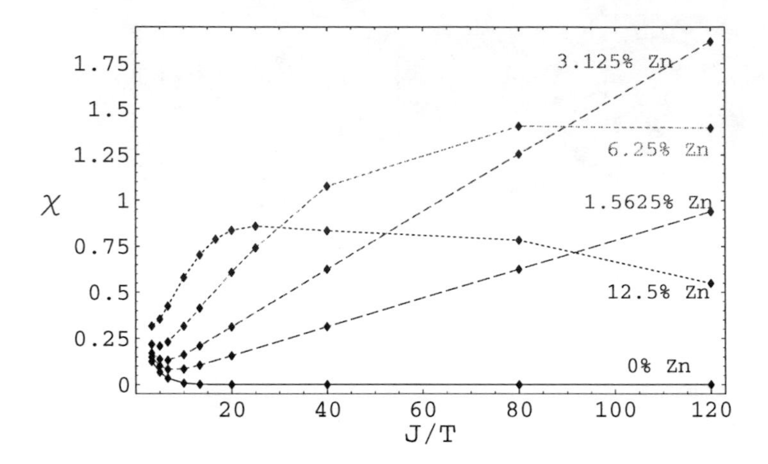

Figure 3.8: Uniform magnetic susceptibility χ as a function of T^{-1} for different regularly distributed impurity concentrations. Up to 3.125%, Zn doping χ is proportional to T^{-1} (Curie-Weiß behavior), higher Zn dopings destroy this behavior.

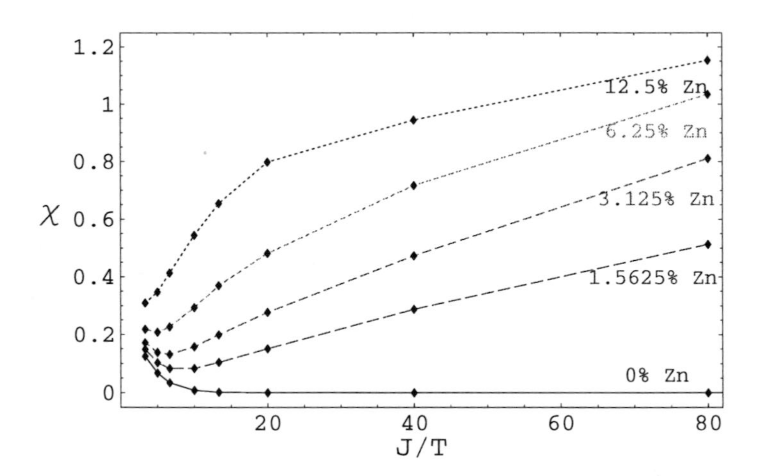

Figure 3.9: Homogeneous magnetic susceptibility χ as a function of T^{-1} for different randomly distributed impurity concentrations. The Curie-Weiß behavior $\chi \propto T^{-1}$ persists up to Zn concentrations of above 10%.

where Δ is the spin gap of the system. The numerical results are in accordance with this form (see Fig. 3.8).
With 2 and 4 impurities, χ can be fitted by

$$\chi(T) = C_0 \frac{e^{-\Delta/T}}{\sqrt{T}} + nC_1 \frac{1}{T}, \tag{3.32}$$

where n is the number of impurities. Hence, the system shows Curie-Weiß behavior. The spin gaps for 0, 2 and 4 impurities are

$$\Delta_0 = 0.501 \pm 0.002 , \quad \Delta_{1.5625\%} = 0.519 \pm 0.005 \quad \text{and} \quad \Delta_{3.125\%} = 0.533 \pm 0.020 ,$$

which is in good accordance with the values of 0.497, 0.526 and 0.553 determined by Iino and Imada for the original Heisenberg ladder [123]. Above a critical Zn doping of about 4%, the Curie-Weiß behavior breaks down. It seems that at dopings below 4% the Zn impurities induce weakly coupled spin-1/2 moments, whereas at higher impurity doping the system behaves more like an effective AF Heisenberg chain of strongly coupled localized spins 1/2 [123].
However, if we insert the impurities at random positions, the Curie behavior persists up to much higher doping concentrations of above 10% (see Fig. 3.9). This can be explained by the fact that with random impurity distributions there will be some very closely neighbored Zn atoms but also some large sections of the ladder without any impurity. These large regions effectively destroy the strong coupling of the induced spin-1/2 moments. The same results have been obtained by Iino and Imada for the original Heisenberg ladder with impurities.
In summary, we have reproduced many features of the AF spin-1/2 Heisenberg ladder with nonmagnetic impurities in our effective bond boson description with Kondo-type couplings to impurity rungs. This shows that the description of the AF regime of clean and impurity-doped cuprate HTSC in terms of triplet quasiparticle excitations is successful. The description presented in this section is a 'prelude' for an SO(5)-symmetric description in which hole-pair bosons are added in order to model also the SC regime of the HTSC (see chapter 5).

4

Computer assisted algebraic manipulations

This chapter describes the implementation and the use of a Mathematica© program for computer aided algebraic manipulations of complex expressions within the formal language of Second Quantization [10]. After a few general remarks on design strategies and on capabilities and limitations of such a software tool, we give some notes on how these ideas can be implemented in Mathematica. Then we present an exemplary application of the new software tool to a current research topic from theoretical solid state physics: we reanalyze the Hubbard-I approximation of the Hubbard model by showing that it is equivalent to an effective Hamiltonian describing fermionic charge fluctuations, which can be solved by a Bogoliubov transformation. As the most important correction in the limit of large U and weak spin correlations, we augment this Hamiltonian by further effective particles, which describe composite objects of a fermionic charge fluctuation and a spin-, density- or η excitation. The scheme is valid both for positive and negative U. We present results for the single particle Green's function for the two dimensional Hubbard model with and without t' and t'' terms, and compare to Quantum Monte-Carlo (QMC) results for the paramagnetic phase. The overall agreement is significantly improved over the conventional Hubbard-I or two-pole approximation. This work has been published in [15].

4.1 Basic strategies

The term manipulation system to be designed here is intended for two kinds of purposes:

- A fully automatic term transformation and simplification (e.g. to prorve that a certain commutator is zero or to solve an eigenvalue problem analytically).
- User-controlled step-by-step transformations of terms from an initial form into a form preferred by the user.

These two aims are somehow contradictory: the first one requires a large set of definitions and replacement rules to be applied automatically whenever possible, while in the second case, at each step only one single replacement rule is to be applied exactly once. Furthermore, modeling the whole formalism of Second Quantization requires such a large set of definitions and rules that the program is likely to be considerably slowed down if all these rules are activated simultaneously. Therefore, a modular and flexible structure of the term manipulation system should be chosen. In the opinion of the author, this is guaranteed by observing the following design principles:

- **formats: input – internal – output**:
  The syntax for reading in new terms from the keyboard should allow a fast and easy input. Two dimensional typesetting with special characters, operators, subscripts and superscripts – which is possible within Mathematica's graphical notebook interface – is not very useful here. The internal storage format should allow compact storage and easy interpretation of the logical structure during computation. The output format should be a compact two dimensional typeset form with subscripts and special characters similar to what the user knows from scientific publications. This considerably helps in interpreting the results and improves efficacity when working interactively with the program.

- **organization of the knowledge base**:
  The set of all known simplification and transformation rules should be grouped into independent modules which can be activated and deactivated separately. So rule sets which are currently not needed can be switched off. This can speed up computation considerably.

- **fixed versus temporary rules**:
  Only clearly simplifying transformations (e.g. the canceling out of two terms) can be applied whenever possible. For all other transformation rules there should be two different application modes: either the rule is applied exactly once and then deactivated again (this temporary mode is useful in interactive transformations), or the rule is automatically applied whenever possible (fixed mode).

- **two-way rules**:
  For many rules, the user should be able to decide in which direction they are to be applied. For example, it might sometimes be useful to replace occupation numbers $n_{i,\sigma}$ by $c^\dagger_{i,\sigma}c_{i,\sigma}$, sometimes one might rather want to simplify $c^\dagger_{i,\sigma}c_{i,\sigma}$ to $n_{i,\sigma}$.

- **extensibility**:
  There should be tools, e.g. templates for defining new objects and rule sets, which support the user in extending the program's capabilities.

4.2 Implementation in Mathematica

The Mathematica package `QM.ma`, which is documented in detail in appendix A.4 of this work, consists of many relatively small packages of rules that can be activated and deactivated independently of each other. Each package can be activated in several ways:

- `On[`*command*`]` activates the definitions associated with the command *command*, i.e. the replacements are automatically performed whenever possible.

- `Off[`*command*`]` deactivates the definitions associated with the command *command*, i.e. the replacements are not performed any more.

- *command*`[...]` or `...//`*command* activates the corresponding definitions as temporary replacement rules (applied only one time, not alwaysi.

User defined rule sets can be added using the routines

- `createRuleSet[`*name*`,`*txt*`,`*incompat*`,`*needed*`,`*unprotec*`,`*rules*`,`*nParam*`:0,`*delayed*`:1]` (define a new package of rules and call it *name*)

- `createCompiledRuleSet[`*name*`,`*txt*`,`*incompat*`,`*needed*`,`*rules*`]` (named package of rules in precompiled form for faster evaluation)

- `createFormat[`*name*`,`*txt*`,`*incompat*`,`*needed*`,`*unprotec*`,`*rules*`]` (named package of output directives for TeX-like typographical output)

with the following argument meanings:

- *txt* is a text message to be issued if the rule set is switched on or off.

- *incompat* is a list containing the rule sets which are incompatible with the present one and which are automatically switched off if this one is activated, whereas

- *needed* is just the contrary: a list of rules to be activated automatically together with the current package.

- *unprotec* is a (possibly empty) list of standard Mathematica commands like `Plus` or `Times` for which new rules are defined in the current set and whose protection tag therefore has to be removed temporarily.

- *rules* is a list of strings "*lhs*`:=`*rhs*" containing the replacement rules forming the present set.

- *nParam* (default: 0) is the number of parameters for parameterized rule sets.

- *delayed* (default: True) controls whether definitions and rules are implemented as 'delayed' (`:=` and `:>`) or 'immediate' (`=` and `->`).

`QM.ma` uses the notation conventions listed in table 4.1 for building up composed expressions of quantum mechanical operators, states, prefactors, numbers and other objects.

In the current version, the fermion operators ($n_{i,\sigma}$, $c^\dagger_{i,\sigma}$ and $c_{i,\sigma}$), the restricted fermion operators $\hat{c}^\dagger_{i,\sigma}$ and $\hat{c}_{i,\sigma}$ from (1.4), the spin operators $\mathbf{S}_i$, S^z_i, S^+_i and S^-_i, operators describing bosonic triplet excitations, and some important operators from SO(5) theory are implemented (a detailed list can be found in appendix A.4). The program knows the commutation and transformation properties of these operators as well as the result when they are applied onto kets and bras in an occupation number representation.

Input form	Output form	Description
`ket[x]`	$\|x\rangle$	q.m. state ("ket")
`bra[x]`	$\langle x\|$	adjoint of q.m. state ("bra")
`ket[0]`, `bra[0]`	$\|0\rangle$, $\langle 0\|$	vacuum state
`bracket[x,y]`	$\langle x, y\rangle$	scalar product ("bra" times "ket")
`commutator[x,y]`	$[x, y]_-$	commutator
`antiCommutator[x,y]`	$[x, y]_+$	anticommutator
`Op1 ** Op2`	*Op1 Op2*	non-commutative multiplication (of operators, kets, bras)
`aj[Op]`	$Op^\dagger$	Hermitean adjunction
`inv[Op]`	Op^{-1}	inversion
`delta[i,j]`	$\delta_{i,j}$	Kronecker-δ
`hBar`	$\hbar$	Planck's constant $\hbar$

Table 4.1: syntax elements for program package `QM.ma`

4.3 Application: Strong coupling theory for the Hubbard model

4.3.1 Introduction

The Hubbard model (see chapters 1 and 2) is the simplest system which presumably incorporates the key features of the strong correlation limit. Understanding this model will be crucial for making any progress with cuprate superconductors, colossal magneto resistance systems or Heavy Fermions. The special problem in this model is that near half-filling it represents a dense system of strongly interacting Fermions, a situation in which a perturbation expansion in U may not be expected to give any meaningful results. In dealing with this model, one can then pursue two opposing strategies: one might expect that despite the strong interaction a perturbation expansion in U remains a meaningful approximation, and apply conventional many-body theory. The latter means that one is treating the kinetic energy exactly, and results in the validity of the Luttinger theorem. On the other hand, by adiabatic continuity this approach will never produce an insulator at half-filling so that we can be sure of its breakdown in the limit of large U/t.
The opposite point of view was taken by Hubbard [11, 24]: in his approximations, the interaction part of H is treated exactly and approximations are made to the kinetic energy. This results in the breakdown of the Luttinger theorem, because the physical electron is effectively split into two particles, one of them corresponding to an electron moving between empty sites, the other to an electron moving between sites occupied by an electron of opposite spin. The energies of formation of these effective particles differ by U, whence for large U the single free-electron band splits up into two bands formed predominantly by the two types of effective particles. Hubbard's approximations, and related schemes like the so-called 2-pole approximations [127–130], have been dismissed by some authors as unphysical, because they do in fact violate the Luttinger theorem away from half-filling.

However, in a recent QMC study for the paramagnetic phase of the Hubbard model [25] we have shown that such criticism is entirely unwarranted: the Fermi surface, if measured in an 'operational way' from the Fermi energy crossings of the quasiparticle band, indeed does violate the Luttinger theorem. The doping dependence of the Fermi surface volume is qualitatively consistent with Hubbard's results and the main discrepancy being the fact that one can rather clearly distinguish 4 'bands' in the spectral functions rather than the 2 bands predicted by the Hubbard-I approximation. More generally, exact diagonalization studies [1] tend to produce for example a photoemission spectrum consisting of a relatively narrow 'quasiparticle band' (which forms the first ionization states) and an 'incoherent continuum'. A simple two-band form of the spectrum, as produced by the Hubbard-I approximation, therefore cannot give a quantitative description of the spectrum. Motivated by these numerical results we have re-examined the Hubbard I approximation and attempted to find the most important corrections to this scheme for the limit of large U and weak spin correlations. We will see that the 4-band structure observed in the QMC simulations can be reproduced quite well by adding two new effective particles which correspond to a composite object of a 'Hubbard quasiparticle' and a spin-, charge-, or η-excitation. These composite objects actually are the quite obvious leading correction over the Hubbard-I approximation, and we will see that including them into the equations of motion leads to an almost quantitative agreement with the numerical results for a variety of different systems.

4.3.2 A reformulation of the Hubbard I approximation

We consider the Hubbard Hamiltonian (1.1) in particle-hole symmetric form, $H = H_t + H_U$ with:

$$\begin{aligned} H_t &= -t \sum_{\langle i,j \rangle} \left(c_{i,\sigma}^\dagger c_{j,\sigma} + \text{h.c.} \right), \\ H_U &= U \sum_i \left(n_{i,\uparrow} - \frac{1}{2} \right) \left(n_{i,\downarrow} - \frac{1}{2} \right). \end{aligned} \tag{4.1}$$

For bipartite lattices one can make use of two symmetry transformations. The first one is the particle-hole transformation: $c_{i,\sigma} \leftrightarrow e^{i\mathbf{Q}\cdot\mathbf{R}_i} c_{i,\sigma}^\dagger$, where $\mathbf{Q} = (\pi, \pi, \ldots, \pi)$. If the kinetic energy contains only nearest neighbor hopping, this transformation leaves the Hamiltonian invariant and exchanges the electron addition spectrum for momentum $\mathbf{k}$ and removal spectrum for momentum $\mathbf{k} + \mathbf{Q}$ at half-filling. Similarly, the transformation $c_{i,\downarrow} \leftrightarrow e^{i\mathbf{Q}\cdot\mathbf{R}_i} c_{i,\downarrow}^\dagger$, $, c_{i\uparrow} \rightarrow c_{i\uparrow}$ inverts the sign of H_U [131]. At half-filling this allows to transform solutions of the positive-U model into those of the negative-U model. This transformation implies that the single-particle spectral function at half-filling is identical for positive and negative U. In table 4.2 some of the operators introduced in this work and their transformation properties under these transformations are listed.
We proceed with the calculation of the single-particle Green's function. As a first step, following Hubbard [11], we split the electron annihilation operator into the two eigenoperators of the interaction part:

$$\begin{aligned} c_{i,\sigma} &= c_{i,\sigma} n_{i,\bar{\sigma}} + c_{i,\sigma}(1 - n_{i,\bar{\sigma}}) \\ &= \hat{d}_{i,\sigma} + \hat{c}_{i,\sigma}. \end{aligned} \tag{4.2}$$

Operator	particle-hole	positive/negative U
$\hat{c}_{i,\uparrow}$	$\mathrm{e}^{\mathrm{i}\mathbf{Q}\cdot\mathbf{R}_i}\hat{d}^{\dagger}_{i,\uparrow}$	$\hat{d}_{i,\uparrow}$
$\hat{c}_{i,\downarrow}$	$\mathrm{e}^{\mathrm{i}\mathbf{Q}\cdot\mathbf{R}_i}\hat{d}^{\dagger}_{i,\downarrow}$	$\mathrm{e}^{\mathrm{i}\mathbf{Q}\cdot\mathbf{R}_i}\hat{c}^{\dagger}_{i,\downarrow}$
$\hat{d}_{i,\uparrow}$	$\mathrm{e}^{\mathrm{i}\mathbf{Q}\cdot\mathbf{R}_i}\hat{c}^{\dagger}_{i,\uparrow}$	$\hat{c}_{i,\uparrow}$
$\hat{d}_{i,\downarrow}$	$\mathrm{e}^{\mathrm{i}\mathbf{Q}\cdot\mathbf{R}_i}\hat{c}^{\dagger}_{i,\downarrow}$	$\mathrm{e}^{\mathrm{i}\mathbf{Q}\cdot\mathbf{R}_i}\hat{d}^{\dagger}_{i,\downarrow}$
S_i^+	$-S_i^-$	$\mathrm{e}^{\mathrm{i}\mathbf{Q}\cdot\mathbf{R}_i}c^{\dagger}_{i,\uparrow}c^{\dagger}_{i,\downarrow}$
S_i^-	$-S_i^+$	$\mathrm{e}^{\mathrm{i}\mathbf{Q}\cdot\mathbf{R}_i}c_{i,\downarrow}c_{i,\uparrow}$
S_i^z	$-S_i^z$	$\frac{1}{2}(n_i-1)$
n_i-1	$1-n_i$	$2S_i^z$
$\hat{C}_{i,j,\uparrow}$	$\mathrm{e}^{\mathrm{i}\mathbf{Q}\cdot\mathbf{R}_i}\hat{C}^{\dagger}_{i,j,\uparrow}$	$-\hat{D}_{i,j,\uparrow}$
$\hat{C}_{i,j,\downarrow}$	$\mathrm{e}^{\mathrm{i}\mathbf{Q}\cdot\mathbf{R}_i}\hat{C}^{\dagger}_{i,j,\downarrow}$	$-\mathrm{e}^{\mathrm{i}\mathbf{Q}\cdot\mathbf{R}_i}\hat{D}^{\dagger}_{i,j,\uparrow}$

Table 4.2: Transformation properties of various operators under particle-hole and positive/negative U transformation.

These obey $[\hat{d}_{i,\sigma}, H_U] = \frac{U}{2}\hat{d}_{i,\sigma}$ and $[\hat{c}_{i,\sigma}, H_U] = -\frac{U}{2}\hat{c}_{i,\sigma}$. Next, we consider the commutators of these 'effective particles' with the kinetic energy. After some algebra, thereby using the identity $n_{i,\sigma}=\frac{n_i}{2}+\sigma S_i^z$, we find:

$$
\begin{aligned}
[\hat{c}_{i,\uparrow}, H_t] &= -t\sum_{j\in N(i)}\Big[(1-\frac{\langle n\rangle}{2})c_{j,\uparrow} + (c_{j,\uparrow}S_i^z + c_{j,\downarrow}S_i^-) - \frac{1}{2}c_{j,\uparrow}(n_i-\langle n\rangle) + c^{\dagger}_{j,\downarrow}c_{i,\downarrow}c_{i,\uparrow}\Big], \\
[\hat{d}_{i,\uparrow}, H_t] &= -t\sum_{j\in N(i)}\Big[\frac{\langle n\rangle}{2}c_{j,\uparrow} - (c_{j,\uparrow}S_i^z + c_{j,\downarrow}S_i^-) + \frac{1}{2}c_{j,\uparrow}(n_i-\langle n\rangle) - c^{\dagger}_{j,\downarrow}c_{i,\downarrow}c_{i,\uparrow}\Big].
\end{aligned} \tag{4.3}
$$

Here, $N(i)$ denotes the z nearest neighbors of site i. Keeping only the first term in each of the square brackets on the r.h.s. (as we will do for the remainder of this section) reproduces the Hubbard I approximation. We specialize to half-filling ($\langle n_{i,\sigma}\rangle = 1/2$) and introduce the Green's functions

$$
G_{\alpha,\beta}(\boldsymbol{k},t) = -\mathrm{i}\,\langle T\alpha^{\dagger}_{\mathbf{k},\sigma}(t)\,\beta_{\mathbf{k},\sigma}\rangle, \tag{4.4}
$$

where $\alpha,\beta \in \{\hat{c},\hat{d}\}$. Then, using the four anticommutator relations $\{\hat{d}^{\dagger}_{i,\sigma},\hat{d}_{i,\sigma}\} = n_{i\bar{\sigma}}$, $\{\hat{c}^{\dagger}_{i,\sigma},\hat{c}_{i,\sigma}\} = (1-n_{i\bar{\sigma}})$, $\{\hat{d}^{\dagger}_{i,\sigma},\hat{c}_{i,\sigma}\} = \{\hat{c}^{\dagger}_{i,\sigma},\hat{d}_{i,\sigma}\} = 0$, we obtain the following equations of motion:

$$
\begin{aligned}
\mathrm{i}\partial_t G_{\hat{c},\hat{c}} &= \frac{1}{2}\delta(t) + \frac{\epsilon_{\mathbf{k}}-U}{2}G_{\hat{c},\hat{c}} + \frac{\epsilon_{\mathbf{k}}}{2}G_{\hat{d},\hat{c}}, \\
\mathrm{i}\partial_t G_{\hat{d},\hat{c}} &= \frac{\epsilon_{\mathbf{k}}}{2}G_{\hat{c},\hat{c}} + \frac{\epsilon_{\mathbf{k}}+U}{2}G_{\hat{d},\hat{c}}, \\
\mathrm{i}\partial_t G_{\hat{c},\hat{d}} &= \frac{\epsilon_{\mathbf{k}}-U}{2}G_{\hat{c},\hat{d}} + \frac{\epsilon_{\mathbf{k}}}{2}G_{\hat{d},\hat{d}}, \\
\mathrm{i}\partial_t G_{\hat{d},\hat{d}} &= \frac{1}{2}\delta(t) + \frac{\epsilon_{\mathbf{k}}}{2}G_{\hat{c},\hat{d}} + \frac{\epsilon_{\mathbf{k}}+U}{2}G_{\hat{d},\hat{d}}.
\end{aligned} \tag{4.5}
$$

Taking into account that the ordinary electron Green's function G and the Green's function called Γ by Hubbard [11] can be written as

$$\begin{aligned} G &= G_{\hat{c},\hat{c}} + G_{\hat{c},\hat{d}} + G_{\hat{d},\hat{c}} + G_{\hat{d},\hat{d}}, \\ \Gamma &= G_{\hat{d},\hat{c}} + G_{\hat{d},\hat{d}}, \end{aligned} \tag{4.6}$$

the resulting equations of motions are precisely those derived in the Hubbard-I approximation:

$$\begin{aligned} \mathrm{i}\partial_t G &= \delta(t) + (\epsilon_{\mathbf{k}} - \frac{U}{2})G + U\Gamma, \\ \mathrm{i}\partial_t \Gamma &= \frac{1}{2}(\delta(t) + \epsilon_{\mathbf{k}} G + U\Gamma). \end{aligned} \tag{4.7}$$

The present formulation, on the other hand, allows for an appealing physical interpretation of the Hubbard-I approximation: we introduce free Fermion operators $h^\dagger_{\mathbf{k},\sigma}$ and $d^\dagger_{\mathbf{k},\sigma}$, which correspond to 'holes' and 'double occupancies'. The Hubbard operators are identified with these as follows:

$$\begin{aligned} \hat{c}_{\mathbf{k},\sigma} &\to \frac{1}{\sqrt{2}} h^\dagger_{-\mathbf{k},\sigma} \\ \hat{d}_{\mathbf{k},\sigma} &\to \frac{1}{\sqrt{2}} d_{\mathbf{k},\sigma}. \end{aligned} \tag{4.8}$$

Then, we can formally obtain the set of equations of motion (4.5) from the following Hamiltonian for the holes and double occupancies:

$$H_{eff} = \sum_{\mathbf{k},\sigma} \Big(\frac{\epsilon_{\mathbf{k}} + U}{2} d^\dagger_{\mathbf{k},\sigma} d_{\mathbf{k},\sigma} - \frac{\epsilon_{\mathbf{k}} - U}{2} h^\dagger_{\mathbf{k},\sigma} h_{\mathbf{k},\sigma} \Big) + \sum_{\mathbf{k}} \Big(\frac{\epsilon_{\mathbf{k}}}{2} d^\dagger_{\mathbf{k},\uparrow} h^\dagger_{-\mathbf{k},\downarrow} + \text{h.c.} \Big). \tag{4.9}$$

The Hamiltonian (4.9) is a quadratic form and readily solved by Bogoliubov transformation:

$$\begin{aligned} \gamma_{-,\mathbf{k},\sigma} &= u_{\mathbf{k}} d_{\mathbf{k},\sigma} + v_{\mathbf{k}} h^\dagger_{-\mathbf{k},\bar{\sigma}}, \\ \gamma_{+,\mathbf{k},\sigma} &= -v_{\mathbf{k}} d_{\mathbf{k},\sigma} + u_{\mathbf{k}} h^\dagger_{-\mathbf{k},\bar{\sigma}}, \end{aligned} \tag{4.10}$$

to yield the familiar dispersion relation

$$E_\pm(\mathbf{k}) = \frac{1}{2} \Big(\epsilon_{\mathbf{k}} \pm \sqrt{\epsilon^2_{\mathbf{k}} + U^2} \Big). \tag{4.11}$$

To compute the spectral weight of the two bands we use $c_{\mathbf{k},\sigma} = \frac{1}{\sqrt{2}}(d_{\mathbf{k},\sigma} + h^\dagger_{-\mathbf{k},\bar{\sigma}})$, whence

$$Z_\pm(\mathbf{k}) = \frac{1}{2}(u_{\mathbf{k}} \mp v_{\mathbf{k}})^2 = \frac{1}{2}\Big(1 \pm \frac{\epsilon_{\mathbf{k}}}{\sqrt{\epsilon^2_{\mathbf{k}} + U^2}}\Big). \tag{4.12}$$

Again, this is the correct Hubbard-I result. The above discussion shows the physical content of the Hubbard-I approximation: the Hamiltonian (4.9) describes Fermionic particle-like and hole-like charge fluctuations, created by $d^\dagger_{\mathbf{k},\sigma}$ and $h^\dagger_{-\mathbf{k},\bar{\sigma}}$, respectively. These 'live' in a background of singly occupied sites. Particle-like and hole-like charge fluctuations are created in pairs on nearest neighbors, and individually can hop between nearest neighbors. The hopping integral for the hole-like particle has opposite sign as that for the electron-like fluctuation as it has to be, and the hopping integrals for both effective particles are $1/2$ times that for the ordinary electrons: this reflects the fact that due to the Pauli principle (say) a spin up electron added to a 'background' of singly occupied sites can propagate to a neighboring site only with a probability of $1/2$ (we are neglecting any spin correlations of the background of singly occupied sites). The factor of $1/\sqrt{2}$ in (4.8) is due to the fact that $\langle \hat{c}_{i,\sigma}^\dagger \hat{c}_{i,\sigma} \rangle = 1/2$. Finally, the particle which stands for the double occupancy has an energy of formation of $U/2$, the hole-like particle has an energy of $-U/2$. As already mentioned, the Hubbard I approximation therefore describes the splitting of the physical electron into two new effective particles which carry with them information on the 'environment' in which the electron has been created. One of them ($d_{\mathbf{k},\sigma}$) moves between sites occupied by an electron of opposite spin, the other one ($h_{-\mathbf{k},\bar{\sigma}}$) between empty sites. This is a quite appealing physical idea, but the above formulation also very clearly highlights the weak points of the Hubbard-I approximation. In addition to the mere truncation of the commutators (4.3), which is an uncontrolled approximation, these are the following: adopting this picture we would have to assume that the states $h_{i,\uparrow}^\dagger d_{j,\uparrow}^\dagger |0\rangle$ and $h_{i,\downarrow}^\dagger d_{j,\downarrow}^\dagger |0\rangle$ are distinguishable (and in fact orthogonal to one another). This, however, is in general not the case: both states have one double occupancy on site j and a hole on site i, and the only difference is that they have been created in different ways. In fact, using (4.8) we find for their overlap

$$\langle\, d_{j,\downarrow} h_{i,\downarrow} h_{i,\uparrow}^\dagger d_{j,\uparrow}\,\rangle = -4\,\langle S_i^- S_j^+ \rangle = -\frac{8}{3}\,\langle \boldsymbol{S}_i \cdot \boldsymbol{S}_j \rangle, \tag{4.13}$$

where we have assumed a rotationally invariant ground state in the second line. The Hubbard-like approximation scheme thus should work only for a state with vanishing spin correlations - we will therefore henceforth assume the spin correlation function $\langle \boldsymbol{S}_i \cdot \boldsymbol{S}_j \rangle$ to be zero.

A second problem is, that the effective Fermions have to obey a kind of hard-core constraint - a site cannot be simultaneously occupied by a hole and a double-occupancy. This constraint is not accounted for in the derivation of the Hubbard-I approximation: the equations of motion are obtained from the Hamiltonian (4.9) by treating the h and d as ordinary free Fermion operators. This problem is the source of the violation of certain sum-rules when the Hubbard-I approximation is applied in the doped case, see the discussion given by Avella *et al.* [132].

One last remark is that the commutators (4.3) are invariant under the particle-hole transformation, i.e. they are transformed into each others Hermitian conjugate. This remains true for the truncated commutators, which give the Hubbard-I approximation, so that the spectral function obtained from this is particle-hole symmetric. This can also be verified directly using (4.11) and (4.12). Finally, the spectral function is manifestly invariant under sign change of U, as it has to be.

4.3.3 Extension of the Hubbard I approximation

We now want to try and derive an improved version of the Hubbard-I approximation. Thereby we will address neither the problem of nonorthogonality of different hole/double-occupancy configurations nor the hard-core constraint - instead, in this work we will restrict ourselves entirely to an approximate way of treating the omitted terms in the commutator relations (4.3). We expect that the present approximation is reasonable for large U (where the density of holes/double occupancies is small whence the hard-core constraint is of little importance) and weak spin correlations (where the nonorthogonality problem is small). Throughout we stick to the case of half-filling and no spin polarization, $\langle n_{i,\sigma}\rangle = 1/2$. We return to the basic commutator relations (4.3) and consider the terms on the r.h.s. which are omitted in the Hubbard-I scheme. The second term in the square bracket is the Clebsch-Gordan contraction of the spin-1/2 operator $c_{j,\sigma}$ and the spin-1 operator $\boldsymbol{S}_i$ into yet another spin-1/2 operator - it describes the coupling of the created hole/annihilated double occupancy to spin excitations. This term may be expected to be the most important one in the limit of large positive U. The third term describes in an analogous way the coupling to density fluctuations, whereas the last term describes the coupling to the so-called η-pair excitation [133]. One may expect that in the case of *negative* U the last two terms are the important ones.

In keeping with the basic idea of the Hubbard approximations, namely to treat the dominant interaction terms exactly, we split also the composite operator into eigenoperators of the U-term and define:

$$\begin{aligned} \hat{C}_{i,j,\uparrow} &= \hat{c}_{j,\uparrow} S_i^z + \hat{c}_{j,\downarrow} S_i^- - \frac{1}{2}\tilde{n}_i \hat{c}_{j,\uparrow} + c_{i,\downarrow} c_{i,\uparrow} \hat{d}_{j,\downarrow}^\dagger, \\ \hat{D}_{i,j,\uparrow} &= \hat{d}_{j,\uparrow} S_i^z + \hat{d}_{j,\downarrow} S_i^- - \frac{1}{2}\tilde{n}_i \hat{d}_{j,\uparrow} + c_{i,\downarrow} c_{i,\uparrow} \hat{c}_{j,\downarrow}^\dagger, \end{aligned} \tag{4.14}$$

where $\tilde{n}_i = n_i - \langle n \rangle$. Under the positive/negative U transformation we have for example $\hat{c}_{j,\uparrow} S_i^z + \hat{c}_{j,\downarrow} S_i^- \rightarrow \frac{1}{2}\tilde{n}_i \hat{d}_{j,\uparrow} + c_{i,\downarrow} c_{i,\uparrow} \hat{c}_{j,\downarrow}^\dagger$, i.e. keeping the at first sight unimportant (for positive U) terms involving density and pairing fluctuations is crucial for maintaining the exact symmetry under sign change of U. We then have

$$\begin{aligned} \left[\hat{D}_{i,j,\sigma}, H_U\right] &= \frac{U}{2}\hat{D}_{i,j,\sigma}, \\ \left[\hat{C}_{i,j,\sigma}, H_U\right] &= -\frac{U}{2}\hat{C}_{i,j,\sigma}, \\ \left[\hat{c}_{i,\uparrow}, H_t\right] &= -t \sum_{j \in N(i)} \left[\frac{1}{2} c_{j,\uparrow} + \hat{C}_{i,j,\sigma} + \hat{D}_{i,j,\sigma} \right], \\ \left[\hat{d}_{i,\uparrow}, H_t\right] &= -t \sum_{j \in N(i)} \left[\frac{1}{2} c_{j,\uparrow} - \hat{C}_{i,j,\sigma} - \hat{D}_{i,j,\sigma} \right]. \end{aligned} \tag{4.15}$$

The operators $\hat{C}_{i,j,\sigma}$ and $\hat{D}_{i,j,\sigma}$ may be thought of describing 'composite objects' consisting of a charge fluctuation and a spin-, density- or η-excitation on a nearest neighbor. Ultimately these composite operators carry the quantum numbers of a single electron, i.e. spin 1/2 and charge 1. We also note that under particle-hole transformation

$$\hat{C}_{i,j,\sigma} \rightarrow \mathrm{e}^{\mathrm{i}\mathbf{Q}\cdot\mathbf{R}_i} \hat{D}_{i,j,\sigma}^\dagger \tag{4.16}$$

i.e. the composite particles transform in the same way as the $\hat{c}_{i,\sigma}$ and $\hat{d}_{i,\sigma}$. Moreover, the commutators (4.15) are invariant under particle-hole transformation, i.e. this transformation transforms them into each others Hermitian conjugates.

We now enlarge the set of Green's functions by allowing $\alpha, \beta \in \{\hat{c}, \hat{d}, \hat{C}, \hat{D}\}$ in (4.4); in the language of the 'effective Fermions' this means that we are introducing additional Fermions corresponding to the composite objects. To obtain a closed system of equations of motion for these Green's functions, we need equations for the $G_{\hat{C}\hat{c}}$ and $G_{\hat{D}\hat{c}}$. As a first step we turn to the commutators $[\hat{C}_{i,j,\sigma}, H_t]$ and $[\hat{D}_{i,j,\sigma}, H_t]$. Here we have to distinguish three cases (see Figure 4.1):
a) the hopping term acts along the bond (i,j) and transports the hole back from j to i.
b) it transports the hole even further away from i
c) it transports the spin- density- or η-excitation away from site i.

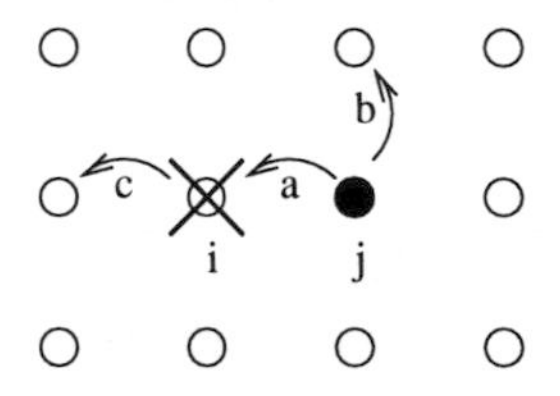

Figure 4.1: Possible hopping processes which couple to $\hat{C}_{i,j,\sigma}$. The cross denotes the spin-, density- or η-excitation, the black dot the hole.

If we want to restrict ourselves to the 4 types of operators $\hat{C}_{i,j,\sigma}$, $D_{i,j,\sigma}$, $\hat{c}_{i,\sigma}$ and $\hat{d}_{i,\sigma}$, we have to neglect the contributions from the processes b and c. These processes would produce 'strings' of excitations along a path of length 2 lattice spacings, and we would have to introduce even more complicated operator products to describe these. Restricting ourselves to processes of the the type a, i.e. replacing $H_t \rightarrow -t\sum_\sigma (c^\dagger_{i,\sigma} c_{j,\sigma} + H.c.)$, straightforward computation gives the surprisingly simple result

$$\begin{aligned} [\hat{C}_{i,j\uparrow}, H_t] &= \frac{3t}{2}\hat{D}_{j,i,\uparrow} + \frac{t}{2}\hat{C}_{j,i,\uparrow} - \frac{3t}{4}(\hat{c}_{i,\uparrow} - \hat{d}_{i,\uparrow}), \\ [\hat{D}_{i,j\uparrow}, H_t] &= \frac{3t}{2}\hat{C}_{j,i,\uparrow} + \frac{t}{2}\hat{D}_{j,i,\uparrow} - \frac{3t}{4}(\hat{c}_{i,\uparrow} - \hat{d}_{i,\uparrow}). \end{aligned} \tag{4.17}$$

Again, these relations are particle-hole invariant, i.e. they are turned into each others Hermitian conjugates by particle-hole transformation. In passing we note that had we reduced the operators $\hat{C}_{i,j\uparrow}$ and $\hat{D}_{i,j\uparrow}$ to comprise only the terms involving spin excitations (as might seem appropriate in the case of large positive U), the commutators would have been much more complicated and in fact the 'Hamilton matrix' H_k to be defined below would have been non-Hermitian.
Next, we need the anticommutators

$$\begin{aligned} \{\hat{C}_{i,j,\uparrow}, \hat{c}^\dagger_{l,\uparrow}\} &= \delta_{j,l}\left[S^z_i S^z_j + S^-_i S^+_j + \frac{\tilde{n}_i \tilde{n}_j}{4} + c^\dagger_{j,\uparrow} c^\dagger_{j,\downarrow} c_{i,\uparrow} c_{i,\downarrow}\right] \\ &+ \delta_{j,l}\left[\frac{1}{2}S^z_i - \frac{1}{4}\tilde{n}_i - \frac{1}{2}(\tilde{n}_i S^z_j + \tilde{n}_j S^z_i)\right] + \delta_{i,l}\left[\hat{c}_{j,\downarrow}\hat{c}^\dagger_{i,\downarrow} - \hat{d}_{i,\downarrow}\hat{d}^\dagger_{j,\downarrow}\right] \\ \{\hat{D}_{i,j,\uparrow}, \hat{c}^\dagger_{l,\uparrow}\} &= \delta_{i,l}\left[\hat{c}_{j,\downarrow}\hat{d}^\dagger_{i,\downarrow} + \hat{d}^\dagger_{j,\downarrow}\hat{c}_{i,\downarrow}\right] \end{aligned} \tag{4.18}$$

Taking the expectation value in the ground state, most of the terms vanish on the basis of symmetries: $\hat{c}_{j,\downarrow}\hat{d}^\dagger_{i,\downarrow} + \hat{d}^\dagger_{j,\downarrow}\hat{c}_{i,\downarrow}$ vanishes due to inversion symmetry of the ground state,

$\hat{c}_{j,\downarrow}\hat{c}^\dagger_{i,\downarrow} - \hat{d}_{i,\downarrow}\hat{d}^\dagger_{j,\downarrow}$ and $\tilde{n}_i$ vanish due to particle-hole symmetry at half-filling. All terms containing unpaired spin operators vanish if we assume that the ground state is invariant under spin rotations (which excludes ferro- or antiferromagnetic solutions). Finally we obtain:

$$
\begin{aligned}
\langle\{\hat{C}_{i,j,\uparrow}, \hat{c}^\dagger_{l,\uparrow}\}\rangle &= \delta_{j,l}\langle \boldsymbol{S}_i \cdot \boldsymbol{S}_j + \frac{\tilde{n}_i\tilde{n}_j}{4} + c^\dagger_{i,\uparrow}c^\dagger_{i,\downarrow}c_{j,\uparrow}c_{j,\downarrow}\rangle, \\
\langle\{\hat{D}_{i,j,\uparrow}, \hat{c}^\dagger_{l,\uparrow}\}\rangle &= 0.
\end{aligned}
\tag{4.19}
$$

It is is easy to see that the expressions whose expectation values are taken are invariant under particle-hole transformation; under the positive/negative U transformation we have:

$$
\begin{aligned}
\frac{\tilde{n}_i\tilde{n}_j}{4} &\leftrightarrow S^z_i S^z_j, \\
c^\dagger_{i,\uparrow}c^\dagger_{i,\downarrow}c_{j,\uparrow}c_{j,\downarrow} &\leftrightarrow S^+_i S^-_j.
\end{aligned}
\tag{4.20}
$$

The expectation value of the anticommutator would be invariant under this symmetry between positive and negative U.

For large positive U the terms $\frac{\tilde{n}_i\tilde{n}_j}{4}$ and ${}^\dagger_{i,\uparrow}c^\dagger_{i,\downarrow}c_{j,\uparrow}c_{j,\downarrow}$ have a negligible expectation value and the only important term comes from the spin correlation. In keeping with our above remarks concerning the role of spin correlations in the Hubbard I approximation we will henceforth take the r.h.s. of (4.19) to be zero. As was discussed above, the Hubbard-I approximation implicitly assumes $\langle \boldsymbol{S}_i \cdot \boldsymbol{S}_j\rangle = 0$, and we will therefore keep this value also in (4.19). We will discuss the consequences of not making this approximation later on.

Using the above commutators and (expectation values of) anticommutators we are now in a position to set up a closed system of equations of motion. In the following we give explicit formulas only for a 1D chain with nearest-neighbor hopping, but the generalization to higher dimensions and/or longer range hopping integrals will be self-evident. We introduce the Fourier transforms

$$
\hat{C}_{\pm,\sigma}(\mathbf{k}) = \sqrt{\frac{4}{3N}} \sum_j \mathrm{e}^{\mathrm{i}\boldsymbol{k}\cdot\boldsymbol{R}_j} \hat{C}_{j,j\pm 1,\sigma},
$$

(and analogously for the $\hat{D}$'s) and define the vector

$$
\boldsymbol{G}_c = \left(G_{\hat{c}\hat{c}}, G_{\hat{d}\hat{c}}, G_{\hat{C}+,\hat{c}}, G_{\hat{C}-,\hat{c}}, G_{\hat{D}+,\hat{c}}, G_{\hat{D}-,\hat{c}}\right).
$$

Here $\hat{C}\pm$ is shorthand for $\hat{C}_{\pm,\sigma}(\mathbf{k})$. Combining (4.3), (4.14) and (4.17), and performing a spatial Fourier transformation the equations of motion are readily found to be

$$
(\mathrm{i}\partial_t - H_k)\boldsymbol{G}_c = \delta(t)B_c \tag{4.21}
$$

where the Hermitian matrix H_k is given by

$$
H_k = \begin{pmatrix}
\frac{\epsilon_{\mathbf{k}}-U}{2} & , & \frac{\epsilon_{\mathbf{k}}}{2} & , & -\tilde{t}\mathrm{e}^{-\mathrm{i}k_x/2} & , & -\tilde{t}\mathrm{e}^{\mathrm{i}k_x/2} & , & -\tilde{t}\mathrm{e}^{-\mathrm{i}k_x/2} & , & -\tilde{t}\mathrm{e}^{\mathrm{i}k_x/2} \\
\frac{\epsilon_{\mathbf{k}}}{2} & , & \frac{\epsilon_{\mathbf{k}}+U}{2} & , & \tilde{t}\mathrm{e}^{-\mathrm{i}k_x/2} & , & \tilde{t}\mathrm{e}^{\mathrm{i}k_x/2} & , & \tilde{t}\mathrm{e}^{-\mathrm{i}k_x/2} & , & \tilde{t}\mathrm{e}^{\mathrm{i}k_x/2} \\
-\tilde{t}\mathrm{e}^{\mathrm{i}k_x/2} & , & \tilde{t}\mathrm{e}^{\mathrm{i}k_x/2} & , & -\frac{U}{2} & , & \frac{t}{2} & , & 0 & , & \frac{3t}{2} \\
-\tilde{t}\mathrm{e}^{-\mathrm{i}k_x/2} & , & \tilde{t}\mathrm{e}^{-\mathrm{i}k_x/2} & , & \frac{t}{2} & , & -\frac{U}{2} & , & \frac{3t}{2} & , & 0 \\
-\tilde{t}\mathrm{e}^{\mathrm{i}k_x/2} & , & \tilde{t}\mathrm{e}^{\mathrm{i}k_x/2} & , & 0 & , & \frac{3t}{2} & , & \frac{U}{2} & , & \frac{t}{2} \\
-\tilde{t}\mathrm{e}^{-\mathrm{i}k_x/2} & , & \tilde{t}\mathrm{e}^{-\mathrm{i}k_x/2} & , & \frac{3t}{2} & , & 0 & , & \frac{t}{2} & , & -\frac{U}{2}
\end{pmatrix}
$$

and $B_c = (\frac{1}{2}, 0, 0, 0, 0, 0)$. The equation system (4.21) can be solved for each momentum and frequency by using the spectral resolution of the Hamilton matrix H_k. This yields the Green's functions $G_{\hat{c}\hat{c}}$ and $G_{\hat{d}\hat{c}}$ for each momentum and frequency. In an analogous way we can also derive an equation system for $G_{\hat{c}\hat{d}}$ and $G_{\hat{d}\hat{d}}$. Thereby the matrix H_k stays unchanged, whereas the r.h.s. is changed into $B_d = (0, 1/2, 0, 0, 0, 0)$. Finally, the full electron Green's function is obtained by adding up the four Green's functions according to (4.6). Upon forming the Laplace transform $G(\mathbf{k}, z)$ we finally obtain the spectral density

$$A(\mathbf{k}, \omega) = -\lim_{\eta \to 0} \frac{1}{\pi} \Im\big(G(\mathbf{k}, \omega + i\eta)\big). \tag{4.22}$$

In passing we note that this way of computing $A(\mathbf{k}, \omega)$ guarantees the validity of the sum rule

$$\int_{-\infty}^{\infty} A(\mathbf{k}, \omega) d\omega = 1. \tag{4.23}$$

Since the particle-hole symmetry of the relations (4.15), (4.17) and (4.19) in turn guarantees particle-hole symmetry of the entire spectral function, we find that the sum-rule for the particle number is fulfilled automatically:

$$\sum_{\mathbf{k}} \int_{-\infty}^{0} A(\mathbf{k}, \omega) d\omega = \frac{N}{2}. \tag{4.24}$$

As a last remark we note that going over to higher dimensions or adding longer ranged hopping integrals poses no problem whatsoever - for each spatial dimension α we have to add four additional rows and columns containing the $\hat{C}_{i,j}$ and $\hat{D}_{i,j}$ where i and j are nearest neighbors in $\pm\alpha$-direction. Similarly, if we add an additional hopping integral t' between 2nd or 3rd nearest neighbors (the number of whom we denote by z') to the Hamiltonian, we have to add $2z'$ rows and columns, containing the $\hat{C}_{i,j}$ and $\hat{D}_{i,j}$ with 2nd or 3rd nearest neighbors i and j. In each case these additional rows and columns contain only mixing terms amongst themselves or with $G_{\hat{c},\hat{c}}$ and $G_{\hat{d}\hat{c}}$, so that the extension is completely trivial.

4.3.4 Comparison with numerics

Following the discussion in the preceding section we can calculate the full electron Green's function, including the (presumably) most important corrections over the Hubbard I approximation in the limit of weak spin correlations and large U. We now proceed to a comparison of the obtained results results for the spectral density $A(\mathbf{k}, \omega)$, with the spectral density obtained from Quantum Monte-Carlo (QMC) simulations. Thereby the temperature for the QMC simulation, $T = 0.33t$, was chosen such that the spin correlation length is only approximately 1.5 lattice spacings - the results thus are probably quite representative for the paramagnetic regime which our approximation aims to describe. Moreover, the value of $U/t = 8$ is already rather large, so that we may also hope to have a small density of holes/double occupancies and the neglect of the hard-core constraint be justified. As a general remark concerning the QMC spectra we note that the MaxEnt procedure used for the analytic continuation to the real axis is most reliable for 'features'

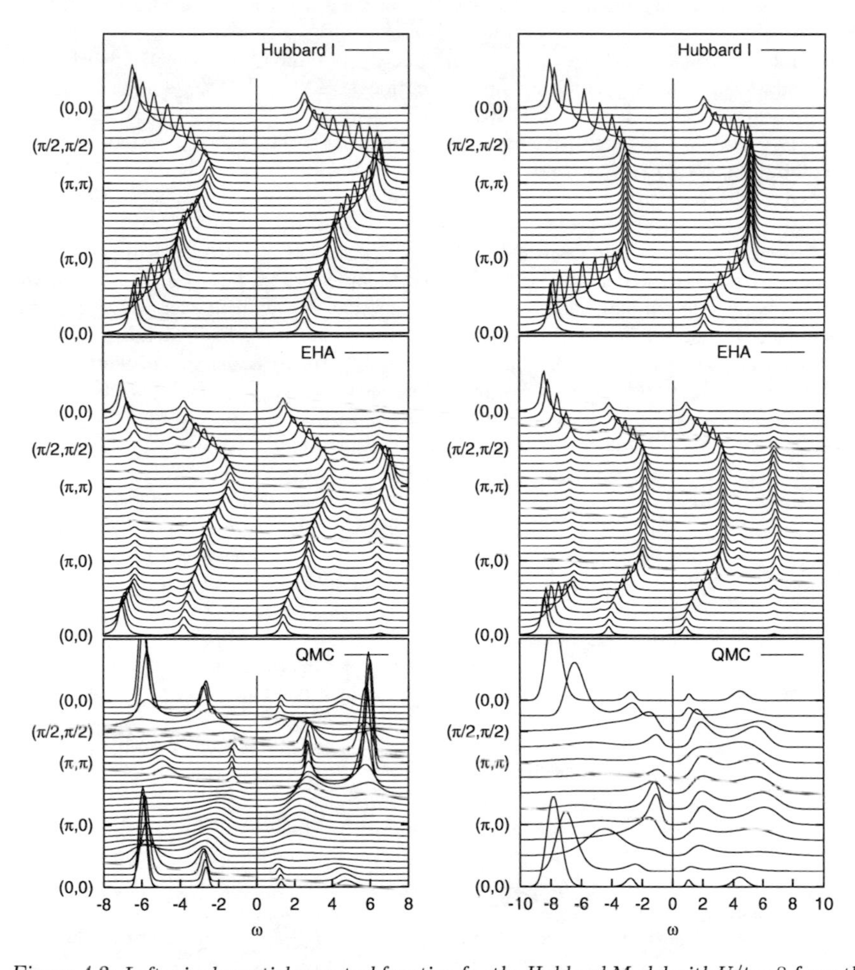

Figure 4.2: Left: single particle spectral function for the Hubbard Model with $U/t = 8$ from the Hubbard I approximation, the extended Hubbard approximation, and QMC simulations on a 20×20 lattice at temperature $T = t/3$. In this as well as in all following figure, the approximate spectra have been given an artificial Lorentzian broadening $\eta = 0.20t$. To compensate for the stronger broadening the QMC spectra have been multiplied by an additional factor of 2.
Right: single particle spectral function for the Hubbard Model with $U/t = 8$, $t' = t/2$ from the Hubbard I approximation, the extended Hubbard approximation, and QMC simulations on an 8×8 lattice at temperature $T = t/3$. To compensate for the stronger broadening the QMC spectra have been multiplied by an additional factor of 4.
(QMC data by C. Gröber and M.G. Zacher, Hubbard I and EHA results by R. Eder)

with large weight - this means that the position of tiny peaks is less accurate than that of large ones.
Figure 4.2 (left) then compares the spectral density obtained from the Hubbard I approximation, our extended Hubbard approximation (EHA) and QMC simulation. The Hubbard I approximation gives only a relatively crude fit to the actual spectral density obtained by QMC. The extended Hubbard approximation, on the other hand, gives an all in all quite correct description of the spectral density. Out of the $10 = 2+4+4$ bands produced by diagonalizing H_k, only four bands do have an appreciable spectral weight. These four intense bands correspond rather well to 4 broad 'bands' of intense spectral weight which can be roughly identified in the QMC result. It is interesting to note that a recent strong-coupling expansion for the Hubbard model by Pairault *et al.* [134] also produced a 4-band structure, although only for the 1D model. The dispersion of the spectral weight along the bands is also reproduced quite satisfactorily be the extended Hubbard approximation. The main difference is the apparently strongly **k**-dependent width of the spectra produced by QMC, which however is outside the scope of the present approximation, which (in the limit $\eta \to 0$) produces sharp δ-peaks without any broadening. A more severe deficiency of the EHA is, that it tends to predict too high excitation energies, resulting in a somewhat too large value of the Hubbard gap. In any way, however, the magnitude of the Hubbard gap comes out better than in the Hubbard-I approximation.
We proceed to the Hubbard model with an additional hopping integral t' between 2nd nearest (i.e. (1,1)-like) neighbors. Figure 4.2 (right) again compares the Hubbard I approximation, the EHA, and the results of a QMC simulation on an 8×8 lattice. The agreement between EHA and the QMC result is again quite good, the main discrepancy being again an overall overestimation of the binding energies. On the other hand, the apparent 4-band structure, the dispersion of the peak energies and the spectral weight agrees well with the numerical result. In particular, the 2 bands in inverse photoemission (i.e. $\omega > 0$) predicted by the EHA can be seen very clearly in the QMC spectra). All in all the agreement is even better than for the case $t' = 0$, which most probably is due to the fact that the spin correlations are weaker with t', so that the assumption of an uncorrelated spin state is better justified in this case.
We proceed to the case of a hopping integral integral t'' between 3rd nearest (i.e. (2,0)-like) neighbors. Here we have chosen $t''/t = 0.25$, because for the larger value $t''/t = 0.5$ the QMC simulation still predicted a metallic state at $U/t = 8$. Figure 4.3 (left) shows the spectral functions for $t''/t = 0.25$. Again, we can roughly identify 4 bands and there is reasonable agreement for the dispersion. The weight of the quasiparticle band in photoemission near (π, π) (at $\omega \approx -2$) is not reproduced very well by the EHA, but again the ubiquitous 4-band structure is rather clearly visible.
Next, we turn to a somewhat indirect check of the approximation. The ordinary electron creation operator is the symmetric combination of the Hubbard operators. However, we might also define the antisymmetric combination

$$\tilde{c}_{i,\sigma} = c_{i,\sigma} n_{i,\bar{\sigma}} - c_{i,\sigma}(1 - n_{i,\bar{\sigma}}). \tag{4.25}$$

This operator has the same quantum numbers as the original electron operator, and therefore obeys the same selection rules. It follows that when acting on the ground state, this operator probes the same final states as the electron operator, the only difference being the matrix element viz. the spectral weight of the respective peak in the spectral

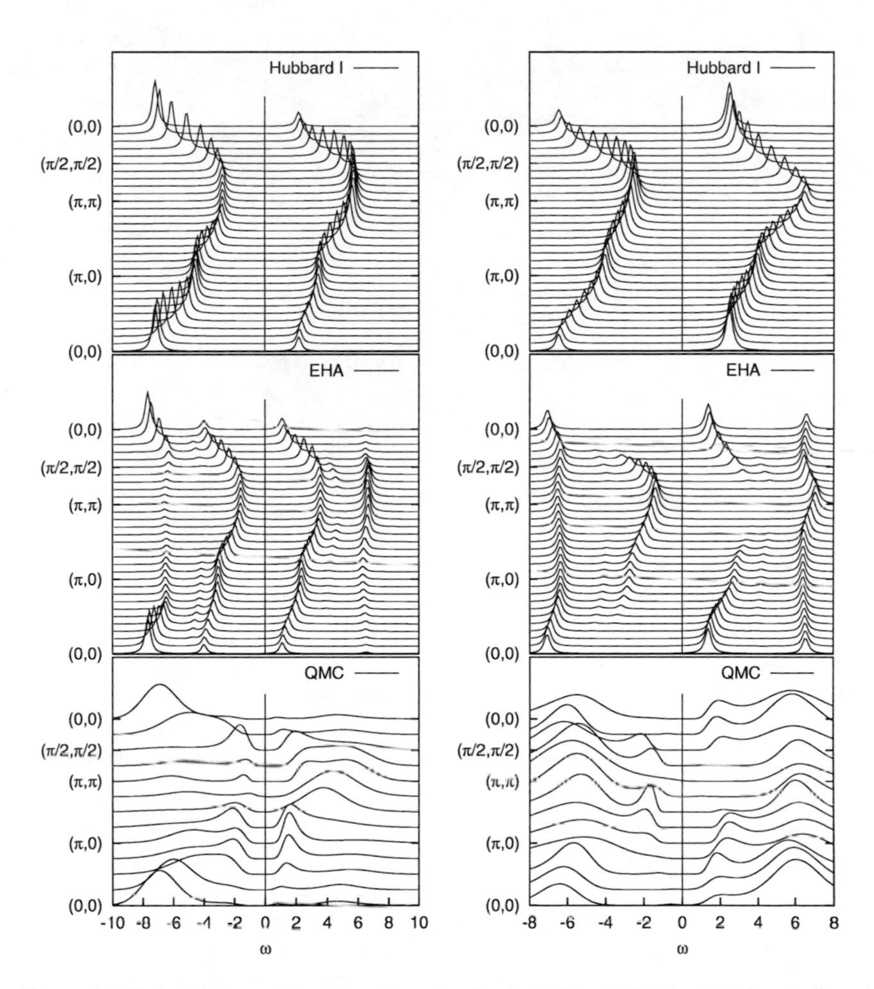

Figure 4.3: Left: single particle spectral function for the Hubbard Model with $U/t = 8$, $t'' = t/4$ from the Hubbard I approximation, the extended Hubbard approximation, and QMC simulations on an 8×8 lattice at temperature $T = t/3$. To compensate for the stronger broadening the QMC spectra have been multiplied by an additional factor of 4.
Right: spectral function of the $\tilde{c}$-operator for the Hubbard Model with $U/t = 8$ from the Hubbard I approximation, the extended Hubbard approximation, and QMC simulations on an 8×8 lattice at temperature $T = t/3$. To compensate for the stronger broadening the QMC spectra have been multiplied by an additional factor of 4.
(QMC data by C. Gröber and M.G. Zacher, Hubbard I and EHA results by R. Eder)

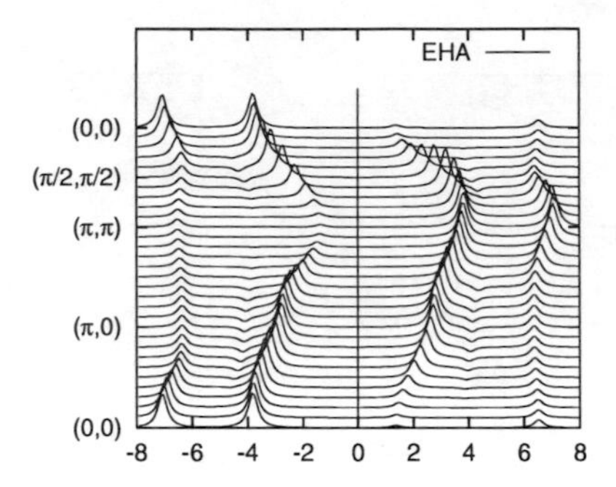

Figure 4.4: Spectral function calculated with the extended Hubbard approximation for $x = -0.2$. (Result by R. Eder)

density. In fact the Green's function

$$\tilde{G}(\mathbf{k}, t) = -i\langle T\tilde{c}^{\dagger}_{\mathbf{k},\sigma}(t)\tilde{c}_{\mathbf{k},\sigma}\rangle \tag{4.26}$$

can be expressed as

$$\tilde{G} = G_{cc} - G_{dc} - G_{cd} + G_{dd}. \tag{4.27}$$

It therefore is easy to calculate within our approximation, and comparison with the QMC result provides an additional check for our description of the electronic structure. Note that the operator $\tilde{c}_{\mathbf{k},\sigma}$ enhances the practically dispersionless bands at $\omega = \pm 6$ - since these bands now have a larger weight, their position and dispersion are more reliable than in the 'ordinary' photoemission spectrum. This is shown in Figure 4.3 (right), where indeed quite good agreement is found between the EHA and the numerical result.
Finally, we turn to the discussion of choosing a nonvanishing expectation value of the anticommutator in (4.19), i.e. we assume that $\langle\{\hat{C}_{i,j,\sigma}, \hat{c}^{\dagger}_{l}\}\rangle = \delta_{j,l}x \neq 0$. This changes the right hand side of the equation system (4.21) to $B_c = (1/2, 0, \sqrt{\frac{4}{3}}x, \ldots)$ but leaves the matrix H_k unchanged. In other words, the *dispersion* of the bands, which is determined by the eigenvalues of H_k stays unchanged and only the spectral weight of the peaks changes. Moreover the sum rules (4.23) and (4.24) retain their validity also in this case. For large positive U and $n = 1$ charge fluctuations will be strongly suppressed and double occupancies will have a small probability, so that the dominant contribution to x comes from the spin correlation function $\langle \boldsymbol{S}_i \cdot \boldsymbol{S}_j\rangle$, whence we should choose $x < 0$. Assuming for example a negative x of moderate value, $x = -0.2$, then leads to little change in the calculated spectral density (see Figure 4.4): the same two bands which had a large spectral weight for $x = 0$ retain a large spectral weight also in this case. There is, however, a rather undesirable feature associated with the bands with small spectral weight: numerical evaluation shows, that for some regions of $\mathbf{k}$-space these bands acquire a small but *negative* weight.
The physical origin of this problem is probably related to nonorthogonalities of basis states: in principle we could interpret the matrix H_k as a Hamiltonian describing (in

2D with only nearest neighbor hopping) 6 types of Fermionic 'effective particles'. Quite generally, the anticommutator-relation $\{a, b^\dagger\} = x \neq 0$ implies that the wave functions corresponding to the Fermi operators $a^\dagger$ and $b^\dagger$ are non-orthogonal. While an exact overlap matrix can never have negative eigenvalues but at most develop zero eigenvalues (indicating that the set of basis states is overcomplete), any approximation to the matrix elements may lead, as an artefact of the approximation, to negative eigenvalues. Since we are using only approximate values for the overlaps, it may happen that we obtain states with a nominally negative norm, whence we can get poles of negative weight. Setting $x = 0$ throughout amounts to assuming that all our effective particles are orthogonal to one another and obviously removes the problem with nonorthogonalities. This seems reasonable, because we are neglecting overlap terms proportional to the spin correlation function already in the Hubbard-I approximation. The lesson then is basically the same as discussed already before: the Hubbard-I approximation is well defined only when applied to an 'ideally paramagnetic' state with no correlations of finite range, and applying it to a state with finite spin correlations represents an approximation.

4.3.5 Conclusion

In summary, we have investigated the most important corrections over the Hubbard-I approximation in the limit $U/t \to \infty$ and electron density $n = 1$. We have seen that the Hubbard-I approximation describes charge fluctuations on a 'background' of singly occupied sites, which is moreover assumed to have zero spin correlations. The charge fluctuations are point-like, and correspond to an electron moving between empty sites and an electron moving between singly occupied sites. We note that a very similar construction can also be applied to the Kondo lattice [135] and in fact reproduces the single particle spectra very well. This is probably due to the fact that the Kondo lattice has a unique ground state in the limit of zero kinetic energy, whereas the ground state of the Hubbard model is highly degenerate in the case $t = 0$.
In our extended scheme for the Hubbard model we have augmented these point-like charge fluctuations by additional 'particles' which are composite in character and consist of a point-like charge fluctuation together with a spin-, density-, or η-excitation. Comparison of the obtained single-particle spectral density with QMC results for a variety of systems showed a quite reasonable agreement. In particular the apparent 4-band structure seen in the numerical spectra finds its natural explanation in the extended Hubbard approximation. We also note that QMC simulations where the spectra of the composite excitations have actually been computed [136], further support the present interpretation. We thus have a quite successful method of computing the full quasiparticle band structure of the Hubbard model, at least in the paramagnetic case and at half-filling.
The present scenario for the nature of the composite excitations also allows to make a connection with various theories for the hole motion in an antiferromagnet [137–142]. There, one is describing holes dressed by antiferromagnetic spin fluctuations. When acting on the Néel state the operators $\hat{C}_{i,j,\sigma}$ obviously describe precisely a hole together with a 'spin wave' on a nearest neighbor or, put another way, a 'string' of length one. The terms which were omitted from the equation of motion for the $\hat{C}_{i,j,\sigma}$ then would correspond to strings of length two and so on. While such longer-ranged strings are apparently of minor importance in the paramagnetic phase, one may expect that they become more

and more important for the description of the dispersion the stronger the antiferromagnetic correlations. The relative importance of such longer ranged strings therefore may be the mechanism for the crossover from the Hubbard-I like dispersion in the paramagnetic phase to the spin-density-wave like dispersion in the antiferromagnetic phase. Similarly, one might think of formulating the entire Hubbard-I approximation also in the antiferromagnetic phase, by constructing the Hamiltonian for charge fluctuations explicitly for a Néel ordered spin background.

5

SO(5) theory of high-temperature superconductivity

This chapter starts with a short overview over basic ideas and mathematical formalism of SO(5) theory of high-temperature superconductivity. We present some successes and weak points of this theory and show that the most important of these weak points are removed by the introduction of a modified, 'projected' SO(5) – or pSO(5) – approach. Finally, we present an effective bosonic pSO(5) model which belongs to the class of four-boson models discussed in chapter 1. The physics of this model will be studied in detail by means of quantum Monte Carlo simulations in the following chapter.

5.1 SO(5) theory

5.1.1 Basic idea

Soon after the discovery of the ceramic cuprate HTSC by Bednorz and Müller [3], it was argued that the physics of the relevant CuO_2 planes could be described by the Hubbard model and its approximation for large Coulomb repulsion, the t-J model with $J \approx 4t^2/U$ [33, 37]. In these models, the electrons tend to condense into local singlet pairs, thereby lowering their energies by up to 100 meV. This gain in energy is due to the magnetic exchange interaction $J \approx 4t^2/U$, for which the commonly accepted values range from 100 to 150 meV [2, 39, 40]. The pairing energy $E = J$ corresponds to a temperature $T^* = E/k_B$ of approximately 1000 K, which is much higher than the experimentally observed critical temperatures for the onset of superconductivity of $T_c \approx 100$ K. This is one of several facts which suggest that in the HTSC pair formation and superconducting phase transition take place on different energy scales and that 'preformed pairs' exist already far above T_c – in contrast to what happens in traditional BCS superconductors [145].

Apart from the superconducting phase (SC), the HTSC possess a second phase with long-range correlations: an antiferromagnetically ordered (AF) phase at hole-doping concentrations near zero (see Fig. 5.1). The transition into the antiferromagnetic state takes place at a Néel temperature T_N which lies somewhere in between T_c and T^*. Typical

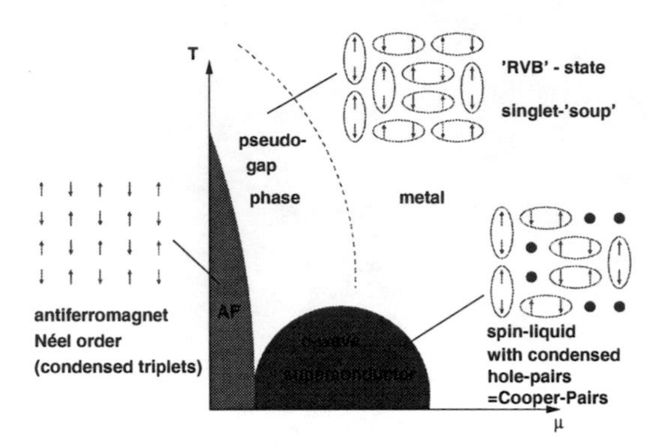

Figure 5.1: Generic $T(\mu)$ phase diagram of the cuprate HTSC. In real HTSC crystals, the chemical potential can be varied by various hole doping concentrations.

values for T_N are 300 to 500 K (see Fig. 5.2). A correct description of the cuprates' low-energy physics must therefore include all three phenomena: long-range AF ordering at temperature T_N, formation of hole pairs at T^* and their condensation into a state with long-range correlations at T_c.
From the theoretical point of view, the t-J model as the simplest effective microscopic model for the cuprates, possesses a surprising variety of possible phases:

- a long-range ordered AF state,
- possibly a d-wave SC state [150],
- a charge density wave (CDW) state [151],
- a spatially homogeneous AF-SC mixed state [163], and
- a disordered 'resonating valence-bond' (RVB) state [37].

Since these different states are not separated by different energy scales, renormalization group methods seem an appropriate tool to construct a unifying effective Hamiltonian capturing the low-energy behavior. However, at present only weak-coupling calculations [152–154] are available. A more promising way is to exploit the symmetries of the low-energy degrees of freedom in order to restrict the set of allowed effective Hamiltonians as much as possible. Starting from these considerations S.C. Zhang proposed the so-called SO(5) theory of HTSC [26]. This theory embeds the U(2) rotational symmetry of spin systems and the U(1) symmetry of charge conservation – both are also obvious symmetries of the t-J and Hubbard model – into a larger five dimensional rotation symmetry, SO(5). SO(5) is the smallest group which contains both U(2) and U(1) as subgroups. The five-dimensional order parameter of this new symmetry group, the so-called superspin, is a combination of the three components of the AF order parameter – the Néel vector (N_x, N_y, N_z) – and the two components forming the order parameter of d-wave SC, namely the real and imaginary part of the superconducting gap function Δ.

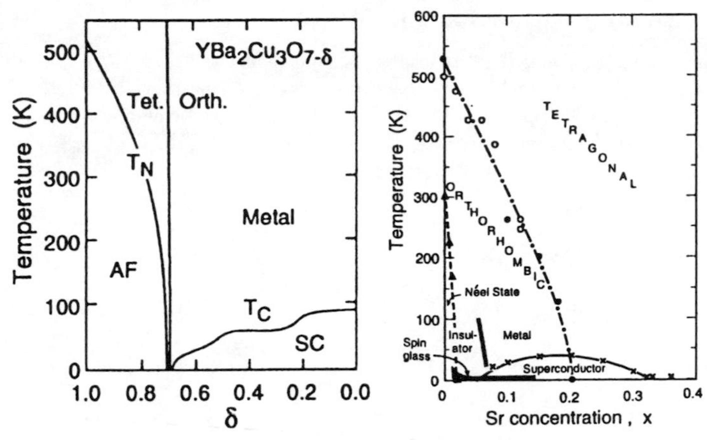

Figure 5.2: Phase diagram of $YBa_2Cu_3O_{6+\delta}$ (left) and $La_{2-x}Sr_xCuO$ (right).

The interpretation of AF and SC phase of the HTSC within SO(5) theory is as follows: in a completely SO(5)-symmetric system the superspin vector can rotate within a five-dimensional sphere, and we would expect mixed states of coexisting AF and SC long-range order. In reality, however, the chemical potential (which controls the hole doping) induces an anisotropy between AF and d-wave SC and explicitly breaks SO(5) symmetry. The chemical potential in SO(5) theory plays the same role as magnetic fields do in SO(3) angular momentum multiplets: a magnetic field splits up these multiplets by introducing small energy offsets between states with different z-spin or L_z quantum numbers (Zeeman effect). Therefore, at half-filling all superspin directions within the 3D 'easy sphere' of zero SC and finite AF order parameter values are energetically favorable, and upon doping the superspin flips into the 'easy plane' of zero AF and finite SC order parameter values [26, 143] (see Fig. 5.3).

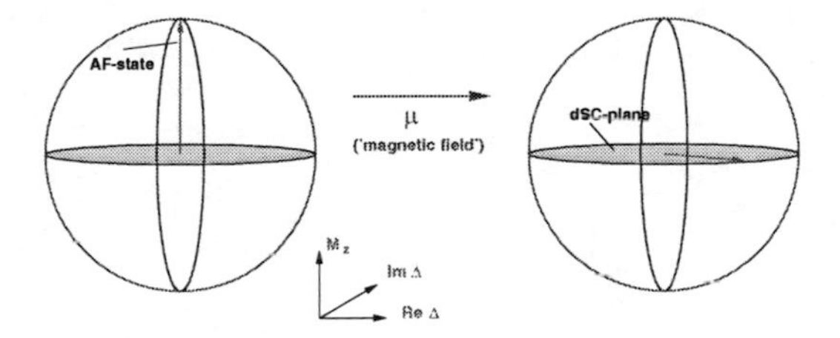

Figure 5.3: The chemical potential forces the superspin into the 'easy sphere' of zero SC order parameter at half-filling and into the 'easy plane' of zero AF order parameter at finite hole doping. (Picture by M.G. Zacher [143]).

In summary, the basic idea of SO(5) theory is that antiferromagnetism and superconductivity are "two faces of the same coin" – just as the electric and magnetic field in the theory of relativity or the proton and the neutron in Heisenberg's isospin concept [157].

5.1.2 Mathematical formalism

Following Ref. [26], [148] and [149], the ideas sketched in the previous section are now cast into a mathematical formalism. Starting point for the construction of SO(5) theory are the AF and d-wave SC phases of the HTSC with their specific symmetries:
AF phase: broken U(2) symmetry:

$$\text{AF order parameter:} \quad N_\alpha = \frac{1}{2} \sum_{\mathbf{k},\sigma,\tau} c^\dagger_{\mathbf{k}+(\pi,\pi),\sigma} \left(\sigma^\alpha\right)_{\sigma\tau} c_{\mathbf{k},\tau}$$

$$\text{generators of U(2):} \quad S_\alpha = \sum_{\mathbf{k},\sigma,\tau} c^\dagger_{\mathbf{k},\sigma} \left(\sigma^\alpha\right)_{\sigma\tau} c_{\mathbf{k},\tau}$$

$$\text{commutation relations:} \quad \left[S_\alpha, S_\beta\right]_- = i\epsilon_{\alpha\beta\gamma} S_\gamma \quad \text{and} \quad \left[S_\alpha, N_\beta\right]_- = i\epsilon_{\alpha,\beta,\gamma} N_\gamma .$$

d-wave SC phase: broken U(1) symmetry:

$$\text{SC order parameter :} \quad \Delta = \frac{\mathrm{i}}{2} \sum_{\mathbf{k},\sigma,\tau} d(\mathbf{k})\, c^\dagger_{\mathbf{k},\sigma} \left(\sigma^y\right)_{\sigma\tau} c_{-\mathbf{k},\tau}$$

with the d-wave prefactor $d(\mathbf{k}) = \cos k_x - \cos k_y$

$$\text{generators:} \quad Q = \frac{N_{el} - N_o}{2}$$

$$\text{commutation relations:} \quad \left[Q, \Delta_i\right]_- = \epsilon_{ij}\Delta_j, \quad \text{where} \quad \Delta_{1/2} = \Re/\Im(\Delta).$$

Here, the σ^α are the *Pauli spin matrices*. The five-dimensional superspin vector $\mathbf{n}$ is then defined as

$$\mathbf{n} = \begin{pmatrix} \Delta_{Re} \\ N_x \\ N_y \\ N_z \\ \Delta_{Im} \end{pmatrix}, \tag{5.1}$$

where N_α are the three components of staggered magnetization, i.e. the AF order parameter, and Δ the superconducting order parameter.
The SO(5) Lie algebra has 10 generators, $L_{ab} = -L_{ba}$, which have the following commutation relations [155]:

$$\left[L_{ab}, L_{cd}\right]_- = i\left(\delta_{ac}L_{bd} - \delta_{ad}L_{bc} - \delta_{bc}L_{ad} + \delta_{bd}L_{ac}\right). \tag{5.2}$$

The U(2)⊗U(1) subgroup contributes four generators, namely the three components of total spin S_x, S_y and S_z as well as the generator of U(1) invariance: the charge pair number (counted from half-filling)

$$Q = \frac{1}{2} \sum_{\mathbf{r}} (n_{\mathbf{r},\uparrow} + n_{\mathbf{r},\downarrow} - 1) .$$

The six new generators have been called 'π operators' by Demler and Zhang [147]; they have the form

$$\pi^\dagger_\alpha := \sum_{\mathbf{k},\sigma,\tau} g(\mathbf{k})\, c^\dagger_{\mathbf{k}+(\pi,\pi),\sigma} \left(\sigma^\alpha \sigma^y\right)_{\sigma\tau} c^\dagger_{-\mathbf{k},\tau} \quad \text{and} \quad \pi_\alpha = \left(\pi^\dagger_\alpha\right)^\dagger \tag{5.3}$$

	spin	charge	momentum
$\Delta_{Re/Im}(\Delta^{\dagger}_{Re/Im})$	0	-1(+1)	0
N_α	1	0	(π,π)
$\pi_\alpha(\pi^{\dagger}_\alpha)$	1	-1(+1)	(π,π)

Table 5.1: Properties of the operators Δ, N_α and π_α

with $\alpha = x, y, z$ and with $g(\mathbf{k})$ being the absolute value of the d-wave prefactor:

$$g(\mathbf{k}) = \mathrm{sign}(\cos k_x - \cos k_y)\,.$$

These four plus six operators, when combined into the 5×5 generator matrix

$$L = \begin{pmatrix} 0 & -(\pi^{\dagger}_x + \pi_x) & -(\pi^{\dagger}_y + \pi_y) & -(\pi^{\dagger}_z + \pi_z) & -Q \\ \pi^{\dagger}_x + \pi_x & 0 & S_z & -S_y & i(\pi^{\dagger}_x - \pi_x) \\ \pi^{\dagger}_y + \pi_y & -S_z & 0 & S_x & i(\pi^{\dagger}_y - \pi_y) \\ \pi^{\dagger}_z + \pi_z & S_y & -S_x & 0 & i(\pi^{\dagger}_z - \pi_z) \\ Q & -i(\pi^{\dagger}_x - \pi_x) & -i(\pi^{\dagger}_y - \pi_y) & -i(\pi^{\dagger}_z - \pi_z) & 0 \end{pmatrix}, \tag{5.4}$$

have the required commutation relations (5.2).
The superspin $\mathbf{n}$ is a vector representation of the SO(5) Lie algebra. It has the following commutation relation with the group's generators L_{ab}:

$$\left[L_{ab}, n_c\right]_- = -i\delta_{bc} n_a + i\delta_{ac} n_b. \tag{5.5}$$

After inserting of the identities $L_{2..4,1} = \pi^{\dagger}_\alpha + \pi_\alpha$ and $n_{2..4} = N_\alpha$ into this equation, we obtain

$$\left[\pi^{\dagger}_\alpha, N_\beta\right]_- = i\delta_{\alpha\beta}\Delta^{\dagger}, \tag{5.6}$$

Hence, the $\pi_\alpha(\pi^{\dagger}_\alpha)$ operators replace triplets by hole pairs or electron pairs, they thus *rotate the AF order parameter into the SC order parameter and vice versa.* More precisely [161]:

$$|\,SC\,\rangle = e^{i\frac{\pi}{2}L_{1\alpha}}|\,AF_\alpha\,\rangle, \tag{5.7}$$

where $|\,AF_\alpha\,\rangle$ is an AF state whose alternating magnetization points α-direction ($\alpha = 2, 3, 4$ corresponds to x, y, z); $|\,SC\,\rangle$ is a SC state with $\langle SC\,|\frac{1}{2}(\Delta^{\dagger} + \Delta)|\,SC\,\rangle \neq 0$. In Table 5.1, we summarize the physical properties of the operators Δ, N_α and π_α, which follow from this 'rotating' effect of the π operators.
Finally, we note that the π operators, unlike Q and $S_{x/y/z}$, do not commute with common microscopic Hamiltonians such as the Hubbard or t-J model. However, the low-energy physics of the Hubbard and t-J model can be described by effective Hamiltonians for which the following approximate eigenoperator equation holds:

$$\left[H, \pi^{\dagger}_\alpha\right]_- \approx \omega_0 \pi^{\dagger}_\alpha \quad \text{with} \quad \omega_0 = \frac{J}{2}\left(1 - \langle n\rangle\right) - 2\mu. \tag{5.8}$$

Here, μ is the chemical potential measured from half-filling $\langle n \rangle = 1$. The validity of this approximation has been checked both in analytical [147] and numerical studies [148, 160] and was further confirmed by renormalization group studies for quite general classes of Hamiltonians [144, 162].

We end this section with the remark that an 's-wave π operator' constructed from $g(\mathbf{k}) = \cos k_x + \cos k_y$ would inevitably fail to fulfill Eq. (5.8). Hence SO(5) theory predicts a d-wave character for the HTSC.

5.1.3 Classification of states: SO(5) multiplets

In SO(3)-symmetric angular momentum algebra, quantum mechanical states can often be classified into multiplets. The square of the total spin of a state and its quantum number $S(S+1)$ determine the multiplet to which the state belongs, and within this multiplet the states are labeled by different z-spin values. From a mathematician's point of view, one thereby assigns the states to irreducible representations (irreps) of SO(3) Lie algebra. In this section a similar classification of SO(5)-symmetric states into irreducible representations of SO(5) Lie algebra will be presented. Further details can be found in [65, 143, 149, 159] and in particular in [148].

First, we define an SO(5)-symmetric generalization of the total spin square operator $\mathbf{S}^2$ known from SO(3)-symmetric systems: the so-called *Casimir-Operator*

$$C = \sum_{a<b}^{5} {L_{ab}}^2. \tag{5.9}$$

From this definition follows that C and H commute, provided that H commutes with Q and S_z and that Eq. (5.8) holds. In this case the eigenvalues of C are good quantum numbers, as well as the eigenvalues of Q and S^z.

Because of the increased complexity of SO(5) with respect to SO(3) we need two different quantum numbers to label the multiplet. We introduce two integer quantum numbers p and q with $p > q$ which will play the role of the quantum number S in SO(3) multiplets. p and q are connected to the eigenvalues ev of the Casimir operator as follows [155]:

$$ev(C) = \frac{1}{2}p^2 + \frac{1}{2}q^2 + q + 2p. \tag{5.10}$$

The dimension of a multiplet is

$$D = (1+q)(1+p-q)\left(1+\frac{1}{2}p\right)\left(1+\frac{1}{3}(p+q)\right). \tag{5.11}$$

SO(5) is a rank-two algebra, therefore two quantum numbers are needed to label the states within a single multiplet. It can be shown that the charge Q and the z-spin S_z are two suitable quantum numbers. The eight remaining generators then induce transitions between different (Q, S_z) states within the multiplets. Figure 5.4 shows the first multiplets of SO(5) with their states plotted in the Q-S^z plane.

The diamond-like (p, p) irreps describe states with even an electron number and therefore integer charge Q. The $(1, 1)$ quintet, for example, contains a hole pair and an electron

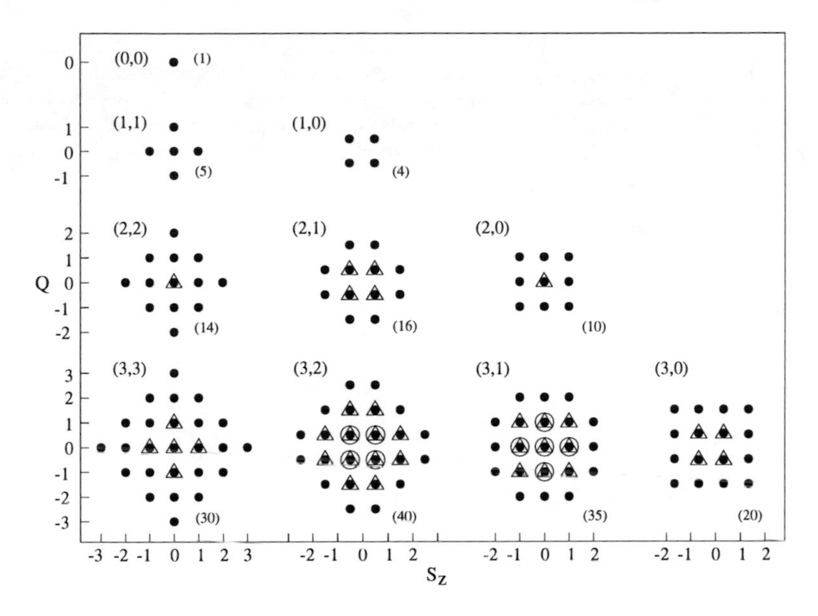

Figure 5.4: The first irreducible representations of SO(5) algebra traced in the Q-S^z plane. Each multiplet is characterized by the quantum numbers (p, q). The single numbers in parentheses are the dimensions of the multiplets. For fixed values of Q and S_z there can be more than one allowed quantum number S, this is symbolized in the figure by superimposing the symbols •, △ and ○.

pair state ($Q = -1, +1$) and three spin excited states or magnons ($Q = 0, S^z = 0, \pm1,$). The lowest energy states of systems with an even number of electrons are in most cases situated in these (p, p) irreps and not in the $(p, p-2)$, $(p, p-4)$, ... multiplets. This is due to the fact that states with given fixed Q, S^z and S values lie in the (p, p), $(p+1, p-1)$, $(p+2, p-2)$, ... multiplets. That means, the larger the difference $p - q$, the bigger the p quantum number of the multiplet. Since the value of p represents the total number of spin and charge excitations in the system, this means that the lowest energy states are grouped in the (p, p) multiplets. This assertion is confirmed by numerical studies of the two-dimensional t-J model [148] which have shown that the model's low energy states can be classified into approximately SO(5)-symmetric (p, p) multiplets. For this practically most relevant special case $p=q$, the equations (5.11) and (5.10) can be written

$$ev(C) = p(p+3) \quad \text{and} \quad D = \frac{(1+p)(2+p)(3+2p)}{6}; \tag{5.12}$$

For the lowest energy states of SO(5)-symmetric systems (with integer charge) this is the equivalent to the equations

$$ev(\mathbf{S}^2) = S(S+1) \quad \text{and} \quad D = 2S+1$$

in SO(3) spin systems.
Besides classifying eigenstates into multiplets, SO(5) symmetry also induces quite restrictive selection rules for allowed transition processes between multiplet states. A detailed derivation and discussion of these selection rules can be found in [65, 143, 149]. Here, we restrict ourselves to giving the results for allowed photoemission and inverse photoemission processes involving only spin-down electrons:

$$\begin{aligned}
&p \rightarrow p \pm 1 \quad \text{and} \quad q \rightarrow q \\
&p \rightarrow p \quad \text{and} \quad q \rightarrow q+1 \\
&Q \rightarrow Q - \frac{1}{2} \\
&S_z \rightarrow S_z + \frac{1}{2} \\
&S \rightarrow S \pm \frac{1}{2}
\end{aligned} \tag{5.13}$$

5.2 Projected SO(5) theory

SO(5) theory of HTSC provides an elegant explanation for many features of the cuprates such as the close vicinity of an AF and a SC phase or the neutron resonance peak at $\mathbf{k}=(\pi,\pi)$ [26, 164–166] and makes a number of experimental predictions [167–171]. Fingerprints of approximate SO(5) symmetry have also been detected in some widely studied microscopic effective models for the CuO_2 planes, e.g. the t-J [148] or the Hubbard model [160]. However, the cuprates' Mott insulating behavior at half-filling severely challenges the validity of SO(5) theory [160, 172–174]: SO(5) symmetry requires collective charge pair excitations to have the same (vanishing) mass as collective spin-wave excitations. The real cuprates, on the contrary, are Mott insulators at half-filling and possess a large energy gap U of several eV due to electron-electron interaction. Therefore, in physically realistic models the 'upper half' of all SO(5) multiplets, i.e. the states with $Q>0$, should be separated from the lower part of the multiplet by the large energy difference U (see Fig. 5.5).
A second fact which cannot be explained within exact SO(5) theory is the gap modulation of the lowest energy peak in angular resolved photoemission spectroscopy experiments (ARPES). Figure 5.6 shows ARPES data for the half-filled HTSC parent compound $Ca_2CuO_2Cl_2$ [175]. The plot reveals clear remnants of the d-wave gap modulation of the superconducting gap in this AF ordered system: The lowest energy peak has minimum excitation energy at $\mathbf{k}=(\pi/2,\pi/2)$ and disperses away from this minimum almost linearly when traced versus $|\cos(k_x a)-\cos(k_y a)|$, reaching its maxima at $\mathbf{k}=(\pi,0)$ and the three equivalent corners of the AF Brillouin zone [143].
In principle, this is another evidence for the strong interrelation of the AF and SC phases in the cuprates and therefore confirms the SO(5) approach. However, in real HTSC materials the modulation amplitude is of order $\Delta^{(AF)}_{mod} \approx J \approx 0.1$ eV in the AF parent compounds, whereas the modulation of the superconducting gap in the doped SC compounds is only of order $\Delta^{(SC)}_{mod} \approx J/10$ (see Fig. 5.7). The original SO(5) theory, which predicts exactly the same gap modulation behavior in the AF and the SC phase, cannot account for these different orders of magnitude.

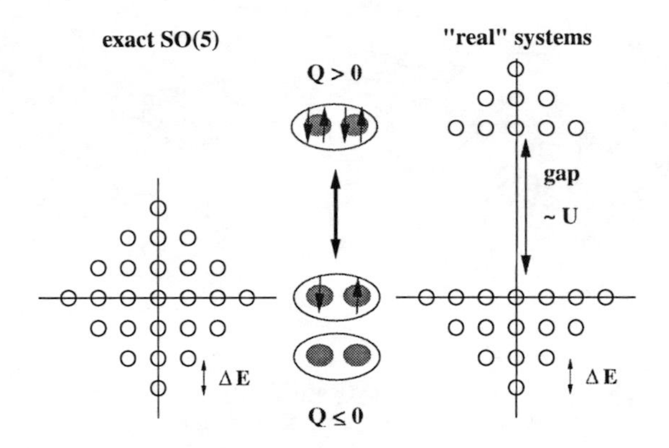

Figure 5.5: In exactly SO(5)-symmetric systems collective charge and spin excitations within an SO(5) multiplet should both have vanishing energy. If exact SO(5)-symmetry is broken by a chemical potential, there are constant energy steps ΔE between neighbored Q levels (left). In physically realistic systems, however, charge excitations above half-filling must pay the large energy U due to the strong on-site repulsion of two electrons in the same orbital.

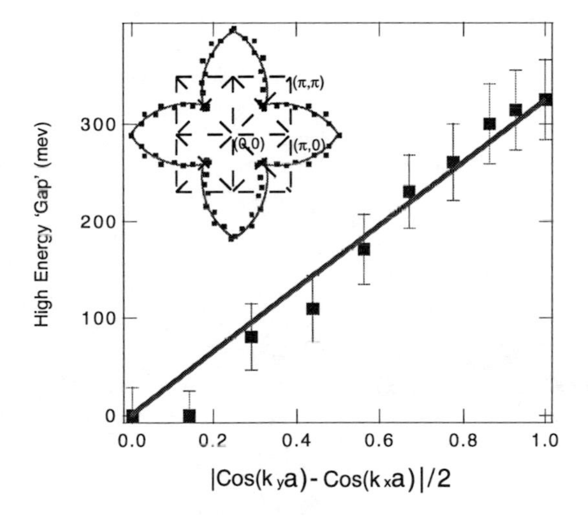

Figure 5.6: ARPES results for $Ca_2CuO_2Cl_2$ from Ref. [175]: the lowest energy peak has minimum excitation energy at $\mathbf{k}=(\pi/2,\pi/2)$ and disperses away from this minimum almost linearly when traced versus $|\cos(k_xa)-\cos(k_ya)|$, reaching its maxima at $\mathbf{k}=(\pi,0)$ and the three equivalent corners of the AF Brillouin zone. This is a d-wave-like dispersion. (picture by M.G. Zacher)

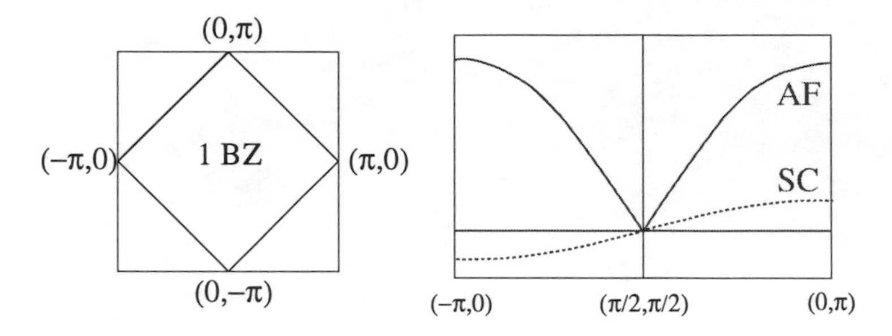

Figure 5.7: Modulation of the superconducting (dotted line) and the antiferromagnetic gap (solid line) along the edge of the AF first Brillouin zone from $\mathbf{k}=(-\pi,0)$ *to* $\mathbf{k}=(0,\pi)$.

The only way to overcome these problems is to incorporate additional SO(5)-symmetry breaking terms into the SO(5) description of the HTSC. Most important is to attain a correct description of the Mott gap at half-filling. This can be done by projecting out all states containing doubly occupied sites (i.e. the $Q > 0$ part of the SO(5) multiplets depicted in Figures 5.4 and 5.5. These states are separated from the states without double occupancies by the energy scale $2U$ of more than 10 eV, which is by orders of magnitude higher than the low-energy scales T_N and T_c. Therefore, states with double occupancies should not even be important as intermediate states for scattering processes. The resulting models, which exactly implement the so-called Gutzwiller constraint of no-double-occupancy, are called 'projected SO(5)' or 'pSO(5)' models.

Figure 5.8 shows one of the central aims when constructing a pSO(5) model: by shifting the chemical potential to the edge of the lower band containing the states without double occupancies, one can at least restore the energetic degeneracy between hole-pair doped and spin excited states within the SO(5) multiplets.

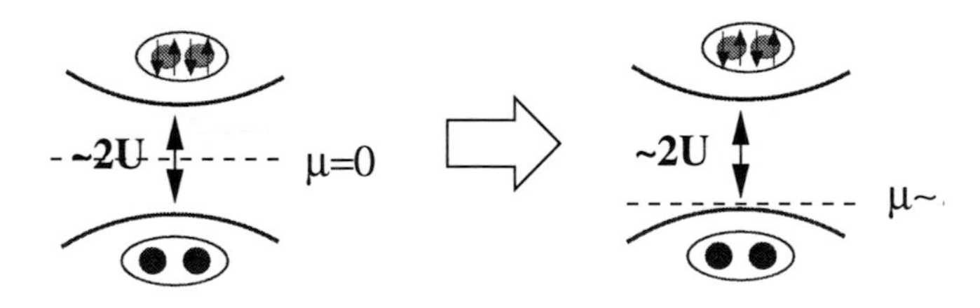

Figure 5.8: A projected SO(5) theory can account for the Mott gap and restore at least the energetic degeneracy of hole-doped and spin-excited states by shifting the chemical potential to the lower band edge. The electron pairs then become irrelevant for the low-energy description of the system since due to the large Mott gap of $2U$.

5.3 A pSO(5)-symmetric effective boson Hamiltonian

5.3.1 Construction of the model

In Ref. [23] a low-energy effective bosonic model has been constructed in which the Gutzwiller constraint of no-double-occupancy is implemented exactly. This is done by projecting out the mode creating particle pair excitations (the so-called π^+ Goldstone mode) and by retaining only the massless magnon and hole-pair modes. In Ref. [149] it has been shown that the low-energy SO(5) excitations on the rung of a ladder can be cast into a picture of 5 hard-core bond bosons: three triplet states ($t^\dagger_{\alpha=2,3,4}$) and particle and hole-pair states ($t^\dagger_p$ and $t^\dagger_h$). As an effective coarse-grained model, this description can be extended to a two-dimensional system, whereby the excitations are now defined on a 2×2 plaquette [23]. The projection is implemented by restricting within the Hilbert space with $t_p(x)=0$. The projected SO(5) Hamiltonian takes the form [23]

$$
\begin{aligned}
H &= \Delta_s \sum_{x,\alpha=2,3,4} t^\dagger_{\alpha(x)} t_{\alpha(x)} + (\Delta_c - 2\mu) \sum_x t^\dagger_{h(x)} t_{h(x)} \\
&- J_s \sum_{<xx'>,\alpha=2,3,4} n_{\alpha(x)} n_{\alpha(x')} - J_c \sum_{<xx'>} (t^\dagger_{h(x)} t_{h(x')} + \text{h.c.}) ,
\end{aligned}
\tag{5.14}
$$

where $n_\alpha = (t_\alpha + t^\dagger_\alpha)/\sqrt{2}$ are the three components of the Néel order parameter. Δ_s and $\Delta_c \sim U$ are the energies to create a magnon and a hole-pair excitation, respectively, at vanishing chemical potential $\mu = 0$. As one can see, the excitation energy for hole pairs can be compensated by μ in order to have equal energies for spin and hole-pair excitations. Due to this partial compensation, the mean-field ground state of this model [23] recovers exact SO(5) invariance at $J_c = 2J_s$ and $\Delta_s = \Delta_c$. However, since the Casimir operator of the SO(5) group does not commute with the Hamiltonian, this invariance is not exact, and a symmetry breaking effect can already be seen at the Gaussian level [23]. Nevertheless, this is the simplest *bosonic* model containing two generic ingredients which are relevant for the high-T_c materials, namely, the Mott gap and the vicinity and possibly common origin of the antiferromagnetic (AF) and of the superconducting (SC) phases.
Interestingly, the Hamiltonian (5.14) turns out to be just a special version of the generic four-boson Hamiltonian (1.15) which we have constructed in chapter 1 as a generic effective bosonic model capturing the low-energy physics of the Heisenberg and the t-J model. Thus, the comparison between (5.14) and (1.15) can be used to relate the parameters J_s, J_c, Δ_c and Δ_c of the pSO(5) model to the physically better understood interaction constants t and J of the t-J model.
In Ref. [23] it was shown that the ground state of the pSO(5) model regains exact SO(5) symmetry at the point

$$
J_c = 2J_s \quad \text{and} \quad \Delta_c - 2\mu = \Delta_s . \tag{5.15}
$$

The modifications of the dynamics with respect to exact S0(5) symmetry can be completely cast into a nontrivial commutation relation between the two SC components of the superspin.

For the *fermionic* sector, a corresponding projected SO(5) model has been shown to provide a natural understanding of the interrelation between SC and AF gaps in cuprate materials [176, 177].

5.3.2 Mean field analysis

The first relevant question concerning the pSO(5) model is whether or not it gives an appropriate description of the phase diagram of the high-T_c cuprates. The second issue is whether the *full* SO(5) symmetry may become exact in the long-wavelength limit in the vicinity of the bicritical point where the AF and SC lines meet [63]. This would imply that antiferromagnetism and superconductivity behave in the same way in the vicinity of this critical point. For a *classical* SO(5)-symmetric model, numerical simulations indicate that the symmetry is indeed restored provided the symmetry-breaking terms have the appropriate sign [30].[1] This is in contrast with the prediction from the ϵ-expansion [179], which would predict a fluctuation-induced first-order transition. This discrepancy clearly indicates that *strong-coupling* effects play an important role and should, therefore, be treated properly. Currently, this can only be done by means of numerical simulations, and we will present such numerical results in the next chapter.
Here, we summarize some results of a mean-field analysis of the pSO(5) model performed in Ref. [23]. Even though the mean-field approximation seems not justified in this system, the obtained results can be used to gain a first insight into the physical properties of the

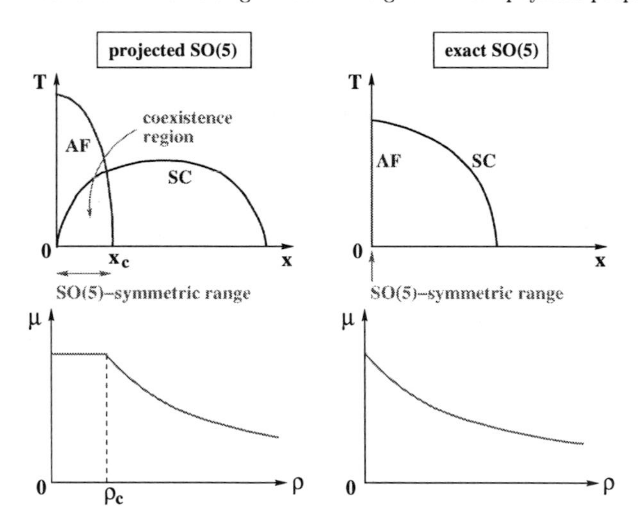

Figure 5.9: Top: AF and SC phases in the pSO(5) model (left) and in exact SO(5) theory (right). Bottom: Doping dependence of the chemical potential in the pSO(5) model (left) and in exact SO(5) theory (right). (The data for the pSO(5) model are mean-field results).

[1] Interestingly, these terms are precisely the ones originating from quantum fluctuations due to the projection, as obtained from a *weak-coupling* expansion [63].

model. Furthermore, they can be used for interesting comparisons with the numerical results which correctly incorporate the strong correlations.
In exactly SO(5)-symmetric models the AF phase only exists at half-filling. On the other hand, the SC phase persists at vanishing doping and even has its maximum T_c there (Fig. 5.9 top right). Both features are in contrast to the real phase diagrams of cuprate HTSC (see Fig. 5.2) in which an AF phase is also observed at small finite hole dopings and the SC phase is pushed away from zero doping. This weak point of exact SO(5) theory is removed in the mean-field physics of the pSO(5) model (Fig. 5.9 top left): here, the AF phase exists at finite doping, maximum T_c is also reached away from zero doping, and there is a mixed phase with both AF and SC long-range order.
The bottom part of Fig. 5.9 compares the doping dependence of the chemical potential in the pSO(5) model and in exact SO(5) theory. The pSO(5) result fits experimental data much better [181] (see Fig. 6.2).
If we exchange the axes in Fig. 5.9 (bottom left), we obtain the hole pair density as a function of the chemical potential in the pSO(5) model. The result is shown in Fig. 5.10: at a critical value μ_c (which is zero for $\Delta_s = \Delta_c$) the hole pair density jumps from zero to the finite value $\rho_c = 0.25$, the points on this jump representing the AF-SC coexistence phase.

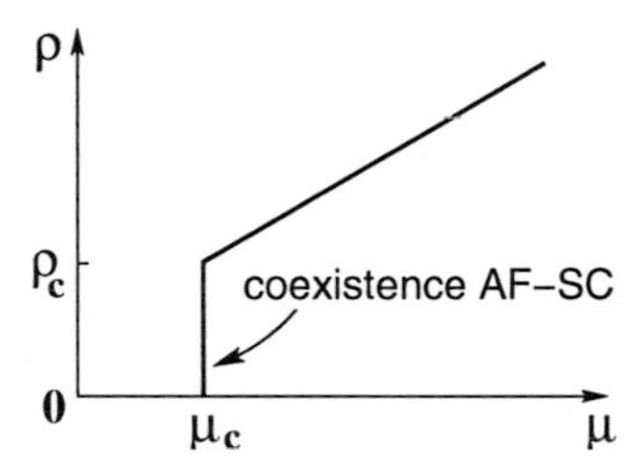

Figure 5.10: hole pair density as a function of the chemical potential in the pSO(5) model (mean-field result). For $\Delta_s = \Delta_c$ the values of μ_c and ρ_c are 0 and 0.25, respectively.

6

Numerical studies of the pSO(5) model

In this chapter, we study the physics of the bosonic pSO(5) model constructed in the previous chapter. As already pointed out, numerical simulations are currently the only methods to study this strongly correlated system in an 'appropriate' manner, i.e. including all particle-particle interaction effects correctly. Our numerical results for two-dimensional (2D) and three-dimensional (3D) lattice geometries were obtained by means of Stochastic Series Expansion (SSE, see chapter 3). We were able to simulate systems of up to 40×40 sites in 2D and $18\times18\times18$ in 3D, including the measurements of arbitrary dynamical response functions. The results show that this model gives a realistic description of the global phase diagram of the high-T_c superconductors and accounts for many of their physical properties. Moreover, we address the question of dynamic restoring of the SO(5) symmetry at certain critical points of the phase diagram.

6.1 The pSO(5) model in two dimensions

6.1.1 Doping dependence of the chemical potential

We start our analysis with numerical simulations on an isotropic 2D square lattice, a good model for the physics of the cuprates' CuO_2 planes. We choose $J_s=J_c/2$, corresponding to the SO(5)-symmetric point in the mean-field approach – see eq. 5.15, and define $J:=J_s$ as our unit of energy. The value of Δ_s turns out not to be crucial for the dynamics of the model or the general structure of the phase diagram, as long as $\Delta_s \lesssim 4J$. We, thus, choose $\Delta_s=J$, and shift the chemical potential so that $\Delta_c=\Delta_s$.
The mean-field results summarized at the end of the previous chapter predict a phase transition from the AF to the SC phase accompanied by a jump in the hole-pair and magnon densities at a chemical potential $\mu_c=0$. In mean field, the system at $\mu=\mu_c$ is in a coherently mixed AF+SC phase and there is no phase separation. However, Gaussian fluctuations change the picture and predict a first-order transition. Here, we want to study this region in more detail with an appropriate *strong-coupling* method. In fact,

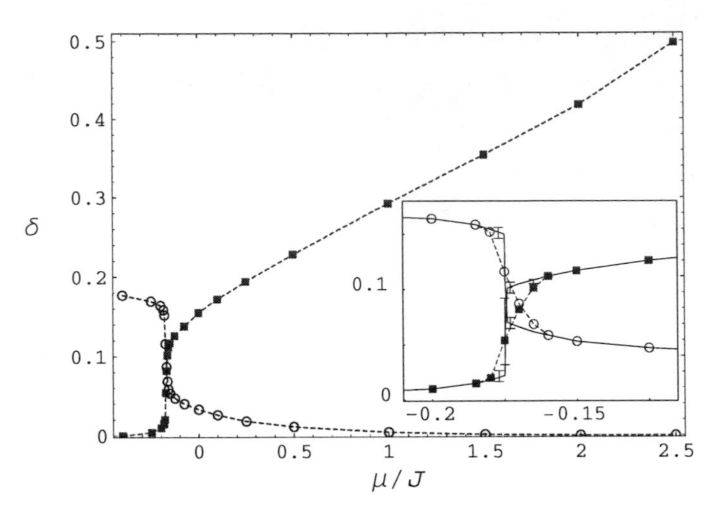

Figure 6.1: Hole concentration $\delta = \rho/2 = \frac{1}{2}\langle t_h^\dagger t_h \rangle$ (filled squares) and magnon density $\sum_\alpha \frac{1}{2}\langle t_\alpha^\dagger t_\alpha \rangle$ (circles) as a function of the chemical potential μ at $T/J = 0.03$. The plotted points result from a finite-size scaling with lattice sizes $V = 10\times 10$, 14×14 and 20×20. The small inlay shows a detailed view to the μ region in which the hole-pair density jumps to a finite value. The additional solid lines with error bars are $T = 0$ data obtained from a simultaneous scaling of $\beta\to\infty$ and $V\to\infty$ (with lattice sizes of $V=8\times 8$, 10×10, 12×12, 16×16, 20×20 and $\beta=4.8, 7.5, 10.8, 19.6$, and 30).

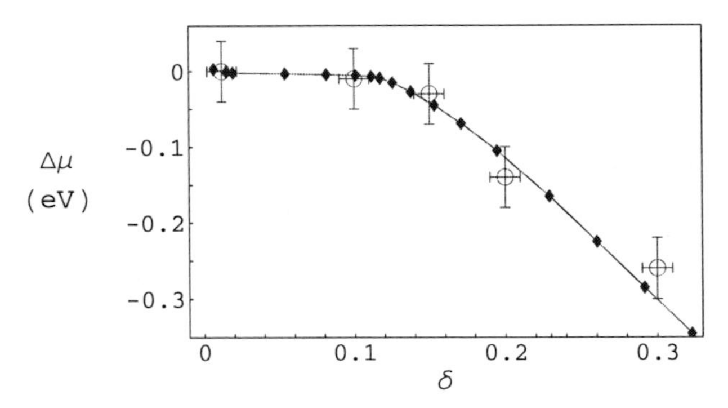

Figure 6.2: Chemical potential shift $\Delta\mu$ as a function of hole concentration δ (open circles) for $La_{2-x}Sr_xCuO_4$ at $T=77$ K. The diamonds and the solid line show numerical data for the infinite lattice obtained by finite-size scaling with $J=220$ meV and $T/J=0.03$, which corresponds to $T\approx 77$ K.

we expect the picture to change appreciably, since no long-range order is allowed in two dimensions.
In Fig. 6.1 we plot the mean hole-pair and magnon densities as a function of the chemical potential for $T/J = 0.03$ and their $T \to 0$ extrapolations. The jump can be clearly seen at $\mu_c = -0.18$, shifted with respect to the classical value, due to the stronger fluctuations of hole pairs as seen in the Gaussian contributions [23]. The size of the jump in the hole-pair concentration $\delta_c \approx 0.11$, as well as as its slope for $\mu > \mu_c$, $\partial\delta/\partial\mu \approx 0.12$, are very close to the mean-field values of 1/8.[1]
Notice that, while the hole-pair density rapidly vanishes below μ_c, a nonvanishing magnon density persists well beyond μ_c even at $T = 0$, due to the fact that the magnon number is not conserved.
A jump in the density as a function of μ has also been seen in $La_{2-x}Sr_xCuO_4$ [181]. A comparison with the experimental data of Ref. [181] is displayed in Fig.6.2. As one can see, our data reproduce the experimental results within error accuracy. The data are fitted by adjusting the energy scale J to 220 meV. This value of J has the correct order of magnitude for a typical magnetic superexchange interaction in the cuprates. However, it should be noted that J is a coarse-grained effective interaction constant and not the microscopic Heisenberg exchange constant J_H. By comparison of Eq. (1.15) and (5.14) these two quantities can be related to each other; the result is $J = 2J_H/3$. That means, our fit corresponds to a Heisenberg interaction constant of $J_H = 330$ meV. This is larger then the commonly accepted value of $J_H \approx 130$ meV but it has the right order of magnitude.

6.1.2 Phase separation and coexistence of AF and SC

The nature of the phase transition at $\mu = \mu_c$ can be determined by studying histograms of the hole-pair distribution for fixed $\mu = \mu_c$. While in an homogeneous phase the density is peaked about its mean value, at $\mu = \mu_c$, we obtain two peaks indicating a first-order transition with a phase separation between (almost) hole-free regions and regions with high hole-pair density (see Fig. 6.3).
In finite lattices, there is always some remaining weight w_i at intermediate densities between the two maxima. with weights W_1 and W_2. However, we have checked in a finite-size scaling of the ratio $R = w_i^2/(W_1\ W_2)$ (using lattice sites up to 32×32) that in the infinite lattice the intermediate densities vanish and the distance between the two peaks remains finite for $T/J < 0.19$. For T/J between 0.19 and 0.2, the intermediate densities rapidly increase, which should indicate the end point of the first-order line. This is confirmed by a plot of the (finite-size scaled) peak positions as a function of temperature displayed in Fig. 6.4.
Interestingly, a phase separation into hole-rich and almost hole-free phases as depicted in Fig. 6.15 can also be observed in real cuprates, for example in the HTSC compound La_2CuO_{4+y} [182]. Fig. 6.5 shows the crystal structure and Fig. 6.6 the phase diagram of La_2CuO_{4+y}. For doping concentrations $y = 0.01$ to 0.055 the material displays a mixture of two phases with different oxygen concentrations: there are alternating AF regions with

[1] The value $\rho_c = 0.25$ given in Fig. 5.10 is the density of hole-pair bosons per coarse-grained effective site of 2×2 microscopic sites, whereas in this chapter we define ρ as the density of holes per microscopic site (copper atom) in order to be consistent with experimentally determined hole densities ρ_{exp}.

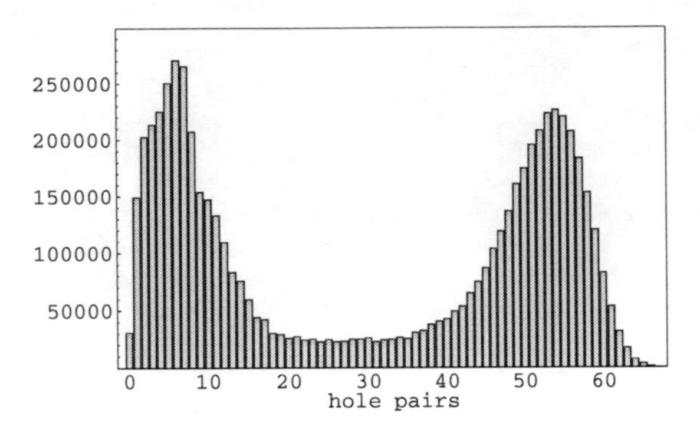

Figure 6.3: Hole-pair density histogram for a 16 × 16 lattice at temperature $T/J = 0.125$ and at the critical chemical potential $\mu_c = -0.35$, recorded in a SSE simulation with 6000000 measurement steps.

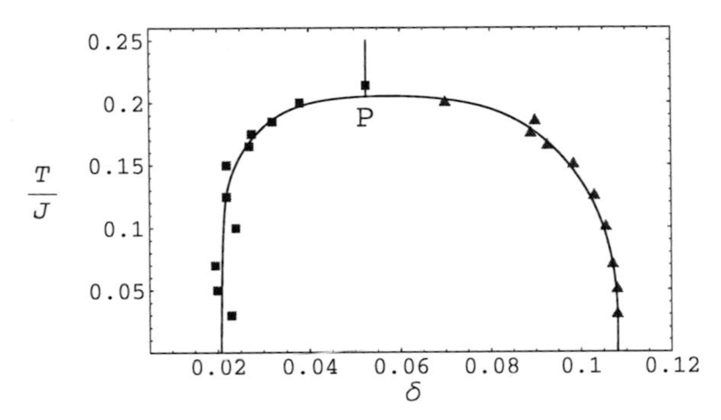

Figure 6.4: Hole densities of the coexisting phases on the first order transition line from (almost) zero to finite hole density at $\mu = \mu_c$ as a function of temperature.

very low doping concentration ($y = 0.01$) and hole-rich 'stage 6' regions ($y = 0.055$) which become superconducting at low temperatures. This is in remarkable accordance with the pSO(5) results in Fig. 6.4 (and also the corresponding diagram for the 3D pSO(5) model, see Fig. 6.15). Due to the fact that each additional oxygen attracts and immobilizes 2 electrons in the CuO_2 planes, thereby introducing 2 holes, the accordance between theory and experiment is perfect even on a quantitative level: the doping densities of $y = 0.01$ and $y = 0.055$ of the two phases exactly correspond to the values $\delta = 2y = 0.02$ and $\delta = 0.11$ obtained from the pSO(5) model.

In strongly interacting electron systems, long-range Coulomb repulsion between the charge carriers should heavily disfavor phase separation. To study the effect of off-site Coulomb interaction we have added additional nearest-neighbor and next-nearest-neighbor Coulomb repulsions V and $V' = 0.67V$ to the pSO(5) model. Fig. 6.7 shows that indeed a relatively moderate Coulomb repulsion of $V/J \approx 0.2$ is enough to completely destroy the phase sep-

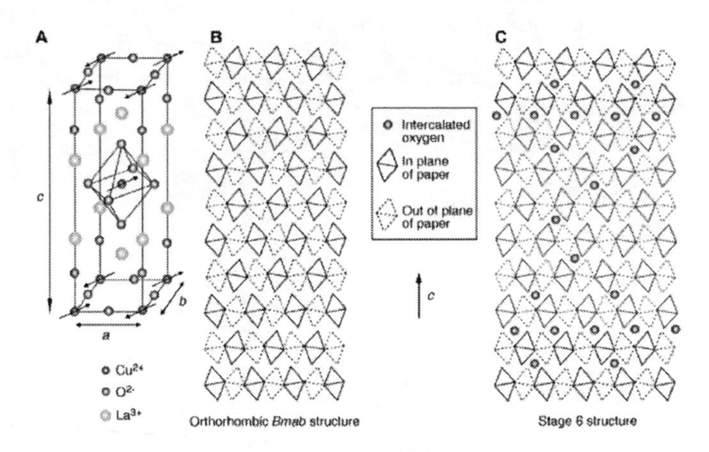

*Figure 6.5: (**A**) The tetragonal unit cell of La_2CuO_{4+y}. (**B**) Schematic crystal structure of undoped La_2CuO_4: the CuO_6 octahedra arrange in a tilted structure. (**C**) Schematic structure of the "stage 6" phase of La_2CuO_{4+y} at oxygen doping $y \approx 0.06$. (Picture by B.O. Wells et al.,* **Science 277**, *p. 1068; reprinted with permission of Science Magazine, Inc. Copyright 1997 American Association for the Advancement of Science).*

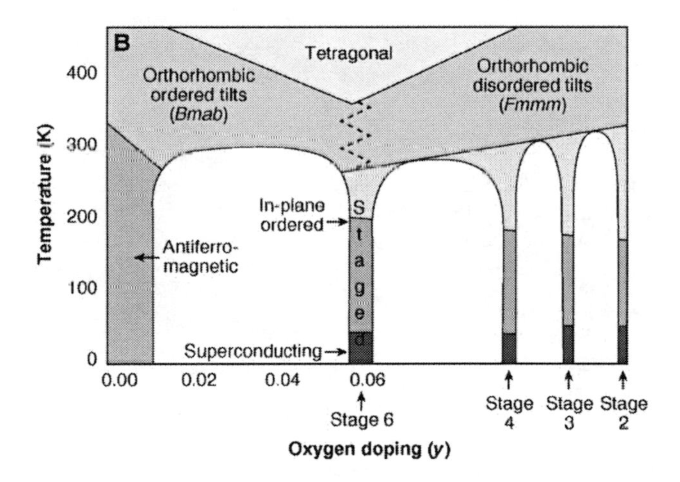

Figure 6.6: Phase diagram of La_2CuO_{4+y} as a function of oxygen doping y. There are several doping ranges in which the material separates into two phases with different doping concentrations, e.g. the doping range $y = 0.01, ..., 0.06$. (Picture by B.O. Wells et al., **Science 277**, *p. 1068; reprinted with permission of Science Magazine, Inc. Copyright 1997 American Association for the Advancement of Science).*

aration. The interesting effect of Coulomb interaction in two dimensions is, thus, to push down the tricritical point into a quantum-critical point at $T=0$. This quantum-critical point now separates the SC from the AF phase and could be a possible candidate for the dynamic restoring of SO(5) symmetry.

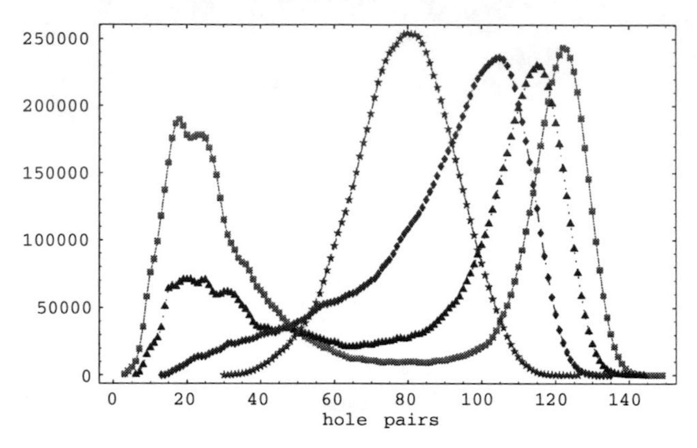

Figure 6.7: Density histograms on 24×24 lattices at $T/J = 0.125$: for a pSO(5) model (squares) showing a pronounced double-peak structure indicating phase separation. The other histograms are obtained by adding a repulsive off-site Coulomb interactions V and $V' = 0.67\,V$ between nearest and next-nearest neighbor hole-pair bosons, respectively, with $V/J = 0.075$ (triangles), $V/J = 0.15$ (diamonds), and $V/J = 0.3$ (stars). As one can see, the Coulomb interaction transforms the structure into one single peak at intermediate densities.

The next question concerns the nature of the phase above the tricritical point (or above the quantum critical point in the presence of Coulomb repulsion): do we find microscopic density modulations in the form of static or dynamic stripes or a phase which is homogeneous also on the microscopic level? Careful studies of the static density-density correlation $\langle n_h(r)\, n_h(0)\rangle$ and the dynamical response function $D(\mathbf{k},\omega) = \langle n_h(\mathbf{k},\omega)\, n_h(\mathbf{k},0)\rangle$ show that there is no sign of stripe formation, neither without nor with a moderate Coulomb repulsion.

6.1.3 Excurse: Kosterlitz-Thouless transitions (KT)

We have anticipated the existence of a SC phase for $\mu > \mu_c$. In fact, in two dimensions at $T>0$ a true long-range order is prohibited by the Mermin-Wagner theorem. However, we can still have a KT [2] phase of finite superfluid density ρ_s at finite temperature, which is identified by a power-law decay of the SC correlation function

$$C_{h(r)} = \left(t^{\dagger}_{h(r)} + t_{h(r)}\right)\left(t^{\dagger}_{h(0)} + t_{h(0)}\right)\,.$$

[2] In recent publications, the KT phase transition is often called Kosterlitz-Thouless-Berezinskii or KTB transition in order to honor the pioneering work of V.L. Berezinskii [183], who published his work three years before J.M. Kosterlitz and D.J. Thouless [184].

The transition separates long-range power-law ($C_{h(r)} \propto r^{-\alpha}$) from rapid exponential decay ($C_{h(r)} \propto e^{-\lambda r}$). A reliable and accurate distinction between these two decay behaviors requires a finite-size scaling with large system sizes, as well as an efficient QMC estimator for the Green functions appearing in the correlation function. With its non-local update scheme and with our new estimators for arbitrary Green functions, SSE provides both. Fig. 6.8 demonstrates how precisely a correlation length can be determined with SSE.

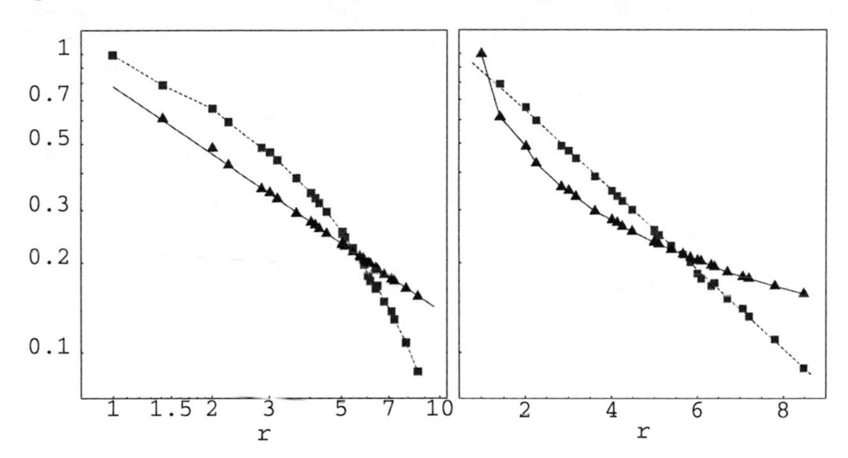

Figure 6.8: Decay behavior of the SC correlation function $C_{h(r)} = \left(t^{\dagger}_{h(r)} + t_{h(r)}\right)\left(t^{\dagger}_{h(0)} + t_{h(0)}\right)$ for $T/J = 0.5$ and for the chemical potentials $\mu = -0.1$ (squares) and $\mu = 0.3$ (triangles). Each point $C_{h(r)}$ was calculated from $C^{(V)}_{h}(r)$ for lattice sizes $V = 8\cdot 8$, $10\cdot 10$, $12\cdot 12$, ..., $24\cdot 24$ by a finite-size scaling (i.e. extrapolation to $C_{h(r)} := C^{(\infty)}_{h}(r)$. The $C_{h(r)}$ data for $\mu = 0.3$ can be fitted by a straight line in a log-log plot of $\log(C_h)$ versus $\log(r)$ (left), the data for $\mu = -0.1$ fit to a straight line in a semi-logarithmic plot of $\log(C_h)$ versus r (right). This proves that the decay is of power-law type at $\mu = 0.3$ and exponential at $\mu = -0.1$, the decay coefficients being equal to the slopes of the two straight lines.

For the point ($T = 0.5J, \mu = -0.1$), which is located in close neighborhood to the phase transition line in fig. 6.9 the semi-logarithmic plot almost perfectly fits a straight line, indicating an exponential decay. For the neighbored points ($T = 0.4J, \mu = -0.1$) or ($T = 0.5J, \mu = 0.1$) we would find the straight fit in a log-log plot instead of the semi-logarithmic plot, which indicates that these points are located in a different phase with power-law decay behavior of $C_{h(r)}$ (see fig. 6.9).
In the numerical simulations presented here, the largest system size used for finite-size scaling was 32×32, in some calculations only 24×24. Earlier QMC studies of simpler Hamiltonians than (5.14), e.g. the work on phase transitions in the quantum XY model (see section 1.1.3) by Ding [185], needed much larger lattice sites to determine phase transition lines and coherence lengths with high precision. However, the QMC methods employed in these works suffer from systematic Trotter discretization errors and from rapidly growing autocorrelation times on large system sizes, which considerably blows up the statistical errors. In Ding's publication [185], for example, typical autocorrelation

times on the largest lattice, the 128×128 lattice, were 5000 update–measurement sweeps. Together with the total sweep number of twice $6 \cdot 10^5$, this means that only about 200 statistically uncorrelated data points were recorded, so that the relative statistical error of the QMC results was about $1/\sqrt{200} \approx 7\%$. In SSE, on the contrary, there are no systematic errors, and the loop update mechanism produces autocorrelation times of the order of 1 even on large lattices. Therefore, the recorded finite-size data typically have relative errors of not more than 10^{-3}. Obviously (see Fig. 6.8), these high precision finite-size data allow for a reliable finite-size scaling even on moderate-sized lattices.
Apart from a change in the decay behavior of the superconducting correlation function, there is a second criterion describing the KT transition point: the superfluid density jumps from zero to a finite value at the KT temperature T_{KT} [186]. Within QMC methods, the superfluid density can be measured quite easily by counting *winding numbers* [187–189]. The vector of winding numbers $\mathbf{W} = (W_x, W_y, W_z)$ describes the behavior of the N particles of a many-particle system if their paths in the world-line representation of the system (see Figures 3.1 to 3.4) are traced from imaginary time $\tau = 0$ to $\tau = \beta$ (which corresponds to the world-line levels $l=0$ and $l=0+L=0$ in Fig. 3.1 to 3.4): W_x (W_y, W_z) is the net number of times the paths of the N particles have wound around the periodic cell in x (y, z) direction in a finite cluster with periodic boundary conditions [187]:

$$\sum_{i=1}^{N}(\mathbf{r}_i^{(\beta)} - \mathbf{r}_i^{(0)}) = W_x L_x + W_y L_y + W_z L_z, \tag{6.1}$$

where L_x (L_y, L_z) are the cluster lengths in x (y, z) direction.
The mean-squared winding number $\langle W_x^2 + W_y^2 + W_z^2 \rangle$ can be related to the so-called *helicity modulus* Υ which describes the free-energy change associated with twisting the order parameter [187]. More precisely, Υ is defined by fixing the order parameter at θ and 0 at opposite ends of a cylinder of area A and length L by imposition of a wall potential and taking [187]

$$\Upsilon = \lim_{A,L\to\infty}\left(-\frac{2L}{\beta\theta^2 A}\ln\left(\frac{\mathcal{Z}(\theta)}{\mathcal{Z}(0)}\right)\right), \tag{6.2}$$

where $\mathcal{Z}$ is the partition function and $\beta = (k_B T)^{-1}$. In Ref. [187] it is shown that the superconducting density ρ_s, helicity Υ and winding numbers $\mathbf{W}$ are related via

$$\rho_s = \left(\frac{m}{\hbar}\right)^2 \Upsilon = \frac{m}{\hbar^2}\frac{\langle \mathbf{W}^2 \rangle L^{2-d}}{d\,\beta} \tag{6.3}$$

with L being the cluster length, which is assumed identical in all d spatial dimensions of the system. For a 2D system, the factor L^{2-d} in (6.3) drops out, and the equation reads

$$\rho_s = \frac{m}{\hbar^2}\frac{\langle \mathbf{W}^2 \rangle}{2\beta}\,. \tag{6.4}$$

In finite-cluster simulations, the jump of ρ_s from zero to a finite value at the KT transition point can be detected more easily than the change in the decay behavior of the correlation function itself.

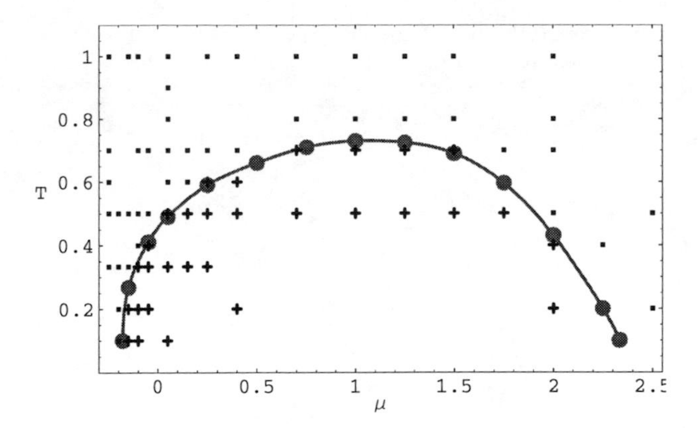

Figure 6.9: Location of the KT phase transition in the projected SO(5) model: The points identify the long-distance decay behavior of the SC correlation function, as obtained by an accurate finite size scaling with lattice volumes $V = 12 \times 14, 14 \times 14, ..., 30 \times 30$ sites. The small squares correspond to an exponential decay, while the crosses indicate a power-law decay. The connected circles trace the transition temperature from the jump in the boson phase stiffness in the $V \to \infty$ limit.

6.1.4 Quasi long-range order and 'superconducting' phase

In this section, we want to search for a KT phase transition line between a phase with quasi long-range order in the hole-hole correlation function – the 2D analogue of a superconducting phase – and a normal conducting phase with short-range correlations. A comparison of the results obtained by the two different criteria to detect a KT transition (see Fig. 6.9) provides a good test for the accuracy of our numerically obtained phase separation line.
Fig. 6.9 plots the phase diagram obtained by applying both criteria independently. The figure shows that the projected SO(5) model indeed has a KT phase with quasi long-range order whose form in μ-T space looks like the one of the high-T_c cuprates. Both criteria result in exactly the same clearly pronounced phase separation line.
It is well known that a similar KT transition cannot occur for antiferromagnets [190], and that the finite-T AF correlation length ξ is always finite and behaves like $\xi \propto e^{2\pi\rho_s/k_BT}$, ρ_s being the spin stiffness. This fact is confirmed by our numerical results.

6.1.5 Spin resonance peak

One of the main features of SO(5) theory is that it provides an elegant explanation for the neutron resonance peak observed in some high-T_c cuprates at $k = (\pi, \pi)$ [26]. Experiments show that the resonance energy ω_{res} is an increasing function of T_c, i.e. ω_{res} increases as a function of doping in the underdoped and decreases in the overdoped region [192]. Here, we address the question whether the T_c dependence of ω_{res} can be reproduced within the projected SO(5) model. To this purpose, we study the spin correlation function at

$k = (\pi, \pi)$. Fig. 6.10 shows the spin correlation spectrum obtained from the projected SO(5) model in two dimensions as a function of the chemical potential.

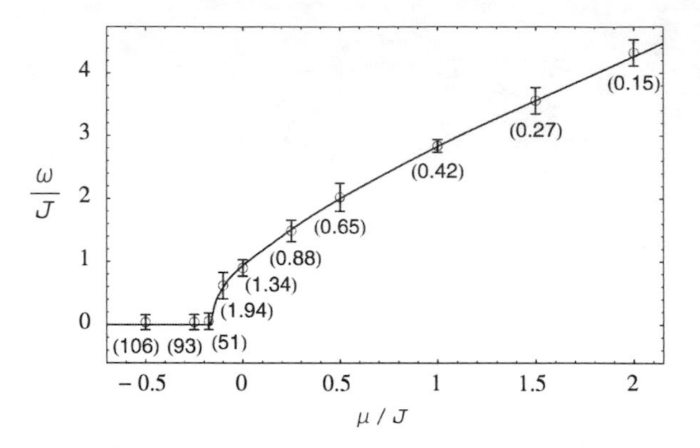

Figure 6.10: Dispersion of the (π, π)-peak of spin correlation as a function of the chemical potential The numbers in parentheses indicate the peak weights, i.e. the area under the peak. (20 × 20 lattice at temperature $T/J = 0.1$).

The magnon-dominated paramagnetic region and the underdoped SC region are described correctly: spin-wave excitations are essentially massless Goldstone modes in the magnon-dominated phase at $\mu < \mu_c$ and become massive when entering into the SC phase. The resonance energy ω_{res} increases monotonically up to optimal doping $\mu_{\text{opt}} \approx 1$. In the overdoped range, however, ω_{res} further increases, in contrast to what happens in the cuprates. This is not surprising given the fact that SO(5) theory was developed to model the interplay of antiferromagnetism and superconductivity in the vicinity of the AF-SC transition, i.e. in the underdoped range. The resonance peak continuously looses weight when increasing μ, which is consistent with experimental observations [192].
A comparison of the critical temperature T_c obtained from Fig. 6.9 and ω_{res} at optimal doping yields the ratio $T_c/\omega_{\text{res,opt}} = 0.23$. This is again in accordance with the corresponding ratio for $YBa_2Cu_3O_{6+x}$, for which the experimentally determined values $T_c = 93\,\text{K}$ (thus $k_B T_C = 8.02$ meV) and $\omega_{\text{res,opt}} = 41$ meV yield $T_c/\omega_{\text{res,opt}} = 0.20$.

6.1.6 How SO(5)-symmetric is the pSO(5) model?

In the preceding paragraphs we have presented some general properties of the pSO(5) model in two dimensions. The question "how SO(5)-symmetric is the pSO(5) model", i.e. the question whether there exists a point on which the full SO(5)-symmetry is dynamically restored, still remains open. As one can see from Eq. (5.14), the excitation energy for hole pairs can be compensated by μ, in order to have equal energies for *local* spin and hole-pair excitations. Due to this partial compensation the mean-field ground state of this model recovers exact SO(5) invariance at $J_c = 2\,J_s$ and $\Delta_s = \Delta_c$ [23]. However, since the Casimir operator of the SO(5) group does not commute with the Hamiltonian, this invariance is not exact, and a symmetry breaking effect can already be seen at the Gaussian level [23].

For a *classical* three-dimensional SO(5)-symmetric model, numerical simulations indicate that the symmetry is asymptotically restored at a bicritical point provided the symmetry-breaking terms have the appropriate sign [30]. This is in contrast with the prediction from the ϵ-expansion [179], which would suggest a fluctuation-induced first-order transition. The discrepancy clearly indicates that *strong-coupling* effects play an important role, calling for an appropriate, i.e. numerically exact treatment, as it is provided by SSE. This assumption is further supported by a recent work by A. Aharony [180].

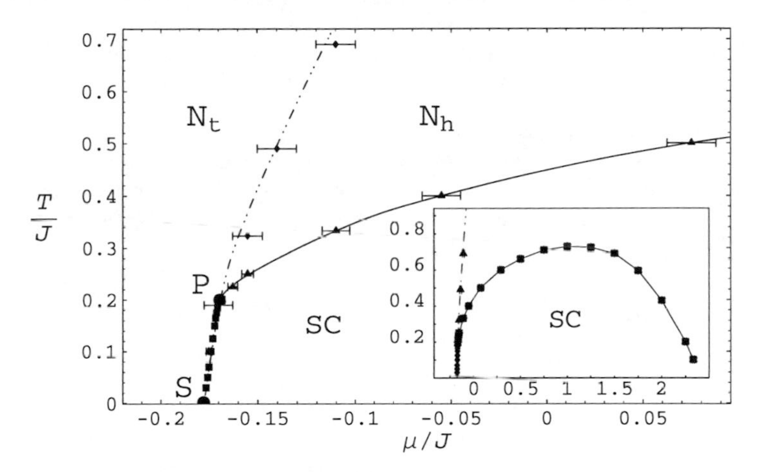

Figure 6.11: Phase diagram of the pSO(5) model: The squares between S and the tricritical point P trace the first-order line of phase separation. The solid line from P to the right edge of the plot traces the Kosterlitz-Thouless transition between the "SC" and the normal state. The dashed line separating N_t (=triplet dominated region) and N_h (=hole pair dominated region) describes the line of equal AF and SC correlation lengths. The small inlay shows the same phase diagram on a larger μ scale, covering the whole "SC" KT phase.

One neccessary condition for an SO(5)-symmetric point is that the formation energies of hole-pair bosons and of magnons are identical. This condition is fulfilled along the line from S to the tricritical point P in Fig. 6.11. Another neccessary condition is that hole pairs and magnons behave in the same way at long distances. This condition is fulfilled on the dashed line in Fig. 6.11, where the AF and SC correlation lengths ξ become equal. Interestingly, these two conditions meet (within error bar accuracy) at the tricritical point P. Of course, here the correlation length is still finite; however, we find relatively large ξ values of order 10 to 15 in the immediate vicinity of point P, demonstrating the importance of SO(5) critical fluctuations in this region.

The dynamic restoring of SO(5) symmetry in the vicinity of the critical point cannot be conclusively answered in $D = 2$ dimensions, even though we have found a critical-point scenario which makes dynamic restoring of SO(5) symmetry at a $T \neq 0$ bicritical point in $D = 3$ a good possibility. For that reason (and because there exists no AF phase at finite temperature in $D = 2$) we proceed to the 3D pSO(5) model in the following section.

6.2 The pSO(5) model in three dimensions

Three aims motivate our studies of the pSO(5) model in three dimensions (3D). First, we expect to find an AF and a SC phase with real long-range order. In particular, we are interested in the interplay between both phases: do we find a coexistence phase of AF and SC as predicted in the mean-field analysis of the pSO(5) model [23] or again a doping range of phase separation as in the 2D case? Second, we want to compare the global phase diagram and the neutron resonance peak at $\mathbf{k}=(\pi,\pi)$ in the 2D and the 3D system. Since the cuprates have a pronounced 2D layer structure with relatively weak couplings between adjacent CuO_2 planes, the 2D and the isotropic 3D model should be two extreme poles for the possible range of properties of real HTSC materials. Third, we want to definitely answer the question whether the pSO(5) model has a certain critical point at which the full SO(5) symmetry is restored. Throughout this section we use the settings $J_s = J_c/2 =: J$ and $\Delta_s = \Delta_c = J$ which correspond to the SO(5)-symmetric point in the mean-field approach [23]. Most numerical data shown in the figures of this section have been obtained by a finite-size scaling with lattice sites $8 \cdot 8 \cdot 8$, $10 \cdot 10 \cdot 10$ and so on up to $16 \cdot 16 \cdot 16$ or $18 \cdot 18 \cdot 18$.

6.2.1 Reference system: a classical 3D SO(5) model

Before discussing the numerical results of the 3D pSO(5) model, we present a classical exactly SO(5)-symmetric Hamiltonian whose phase diagram and scaling behavior have been studied in detail in Ref. [30] by means of classical Monte Carlo simulations (MC). Classical MC are by orders of magnitude easier to perform and less resource demanding than QMC simulations, hence very large system sizes can be simulated and highly accurate data are obtained. More precisely, we want to refer to the Hamiltonian studied in the first part of Hu's work, i.e.

$$H = -J \sum_{\langle i,j \rangle} \left(\mathbf{s}_i^{(AF)} \cdot \mathbf{s}_j^{(AF)} + \mathbf{s}_i^{(SC)} \cdot \mathbf{s}_j^{(SC)} \right) + \mu \sum_i \left(\mathbf{s}_i^{(AF)} \right)^2 , \tag{6.5}$$

where $s_i^{(AF)}$ is the 3D AF order parameter on site i and $s_i^{(SC)}$ the 2D SC order parameter[3] The μ term is thus the only term which breaks SO(5)-symmetry, with $\mu = 0$ being the SO(5)-symmetric point.

Hu established the $T(\mu)$ phase diagram depicted in Fig. 6.12; the model has an AF and a SC phase which meet at a bicritical point $(T_b, \mu_b = 0)$. The phase boundary lines of the AF and the SC phase merge tangentially into the bicritical point, which is an important characteristics of SO(5)-symmetry [30]. The following scaling properties were determined by Hu and will be be used to study the restoring of SO(5)-symmetry in the pSO(5) model:

- Far away from the bicritical point, at $\mu = J/2$, the long-range SC order parameter Υ (the helicity modulus, see section 6.1.3) scales as $\Upsilon \propto \left(1 - T/T_c(\mu)\right)^{\nu}$ with $\nu = 0.666 \pm 0.004$.

- Far into the AF phase, at $\mu = -J/2$, the growth of the AF order parameter $\frac{1}{N}\sum_i (s_i^{(AF)})^2$ follows the law $\frac{1}{N}\sum_i (s_i^{(AF)})^2 \propto \left(1 - T/T_N(\mu)\right)^{\beta_3}$ with $\beta_3 = 0.363 \pm 0.001$.

[3] Ref. [30], writes (6.5) with $g = (\Delta_s - \Delta_c + 2\mu)/2$, which is just μ for our choice of Δ_s and Δ_c.

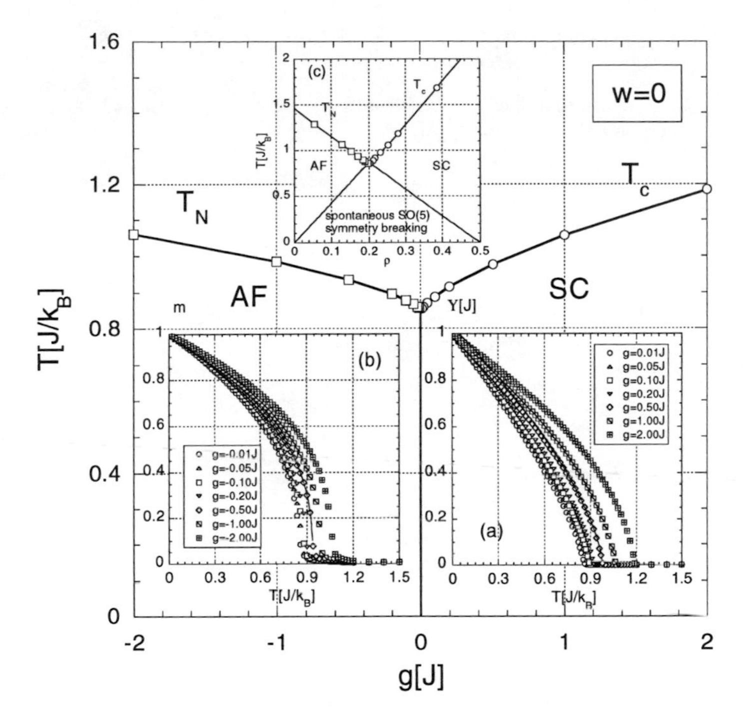

Figure 6.12: Phase diagram of a classical SO(5)-symmetric model with symmetry-breaking chemical potential term $g=(\Delta_s-\Delta_c)/2+\mu$. The big plot shows the $T(g)$ phase diagram, inset c) the $T(\rho)$ phase diagram. Inset a) is the temperature dependence of the helicity modulus for different values of g and inset b) analogously the temperature dependence of the Néel order parameter. (Picture from X. Hu, cond-mat/0011203 v2).

- In the vicinity of the bicritical point, the onsets of both the AF and the SC order parameter become steeper. For an analysis of the crossover phenomenon, we chose an ansatz for the behavior of Υ in the range $T<T_c(\mu)$ and $\mu>0$ which is suggested by scaling theory [30, 194]:

$$\Upsilon(T,\mu)\propto(\mu-\mu_b)^{\nu_5/\phi}\times f\big((T/T_b-1)/(\mu-\mu_b)^{1/\phi}\big)\,. \tag{6.6}$$

Here, ν_5 is the *critical exponent for correlation length* at $n=5$ and ϕ the *crossover exponent.* Using (6.6) the values of ν_5 and ϕ can be determined in two steps. First, performing a μ scan of $\Upsilon(T=T_b,\mu)$ returns the ratio $\nu5/\phi$:

$$\Upsilon(T_b,\mu)/\Upsilon(T_b,\mu')=\big((\mu-\mu_b)/(\mu'-\mu_b)\big)^{\nu_5/\phi}\,. \tag{6.7}$$

Then, ϕ is obtained from the slopes $\frac{\partial}{\partial T}\big(\Upsilon(T,\mu)/\Upsilon(T,\mu')\big)$ via

$$\phi=\ln\left(\frac{\mu_2-\mu_b}{\mu_1-\mu_b}\right)\Big/\ln\left(\frac{\partial}{\partial T}\frac{\Upsilon(T,\mu_1)}{\Upsilon(T,\mu_1')}\Big|_{T=T_b}\Big/\frac{\partial}{\partial T}\frac{\Upsilon(T,\mu_2)}{\Upsilon(T,\mu_2')}\Big|_{T=T_b}\right) \tag{6.8}$$

if μ_1, μ_1', μ_2, and μ_2' are related by $(\mu_1 - \mu_b)/(\mu_1' - \mu_b) = (\mu_2 - \mu_b)/(\mu_2' - \mu_b) > 0$. Hu finds the values $\nu_5/\phi = 0.523 \pm 0.002$ and $\phi = 1.387 \pm 0.030$.

- If the scaling ansatz (6.6) holds, the AF and the SC phase transition line near the bicritical point should be of the form

$$B_2 \cdot (\mu - \mu_b)^{1/\phi} = \frac{T_c(\mu)}{T_b} - 1 \quad \text{and} \quad B_3 \cdot (\mu_b - \mu)^{1/\phi} = \frac{T_N(\mu)}{T_b} - 1\,. \tag{6.9}$$

The ratio B_2/B_3 should be given by the inverse ratio between the AF and SC degrees of freedom, i.e. [195]

$$B_2/B_3 = 3/2\,. \tag{6.10}$$

This is one important possibility to directly measure the number $5 = 3 + 2$ in an SO(5)-symmetric model [196]. The values determined by Hu indeed have the correct ratio: $B_2 = 1/4$ and $B_3 = 1/6$.

6.2.2 Phase diagram

After these preliminaries we proceed to the phase diagram of the 3D pSO(5) model. Figure 6.13 shows that we find an AF and an SC phase as expected. Furthermore, the two phase transition lines merge into a bicritical point (at $T_b = 0.960 \pm 0.005$ and $\mu_b = -0.098 \pm 0.001$) just as in the classical SO(5) system in Fig. 6.12. The line of equal correlation decay of hole-pairs and triplet bosons also merges into this bicritical point P – a neccessary condition for the restoration of SO(5)-symmetry at this point. Unlike the corresponding phase in the classical model, the SC phase extends only over a finite μ range; this is due to the hardcore constraint of the hole-pair bosons and agrees with experimentally determined phase diagrams of the cuprates. Obviously, the quantum mechanical pSO(5) model is 'more physical' in this aspect than the classical SO(5) model.
However, in real cuprates the ratio between the maximum temperatures T_c and T_N is about 0.17 to 0.25 (see Fig. 5.2), whereas in the pSO(5) model we obtain the values $T_c/J = 1.465 \pm 0.008$ at $\mu_{opt}/J \approx 1.7$ and $T_N/J = 1.29 \pm 0.01$ at $\mu \to \infty$, hence T_c is slightly larger than T_N. In section 6.3, we will see how the correct ratio T_c/T_N can be reproduced in a modified pSO(5) model. On the other hand, T_c and T_N very accurately fulfill the equation $(T_c - T_b)/(T_N - T_b) = 3/2$. Together with (6.9) and 6.10, this is a first hint suggesting that a large region around the bicritical point is approximately SO(5)-symmetric.
A closer look to the phase transition line between the points S and P (see Fig. 6.14) reveals that this line is not vertical as in the classical SO(5) model but slightly inclined: At $T = 0$, the critical chemical potential is $\mu_c = -0.148$ and at $T = T_b$ the value is $\mu_c = \mu_b = -0.098$. This indicates that a finite latent heat is connected with the AF-SC phase transition. Moreover, this means that in contrast to the classical model, μ is not a scaling variable for the bicritical point P.

6.2.3 Coexistence of AF and SC and phase separation

By analogy to the 2D pSO(5) model we expect the AF-SC phase transition line below the bicritical temperature T_b to be of first order with a phase separation into a hole-rich SC

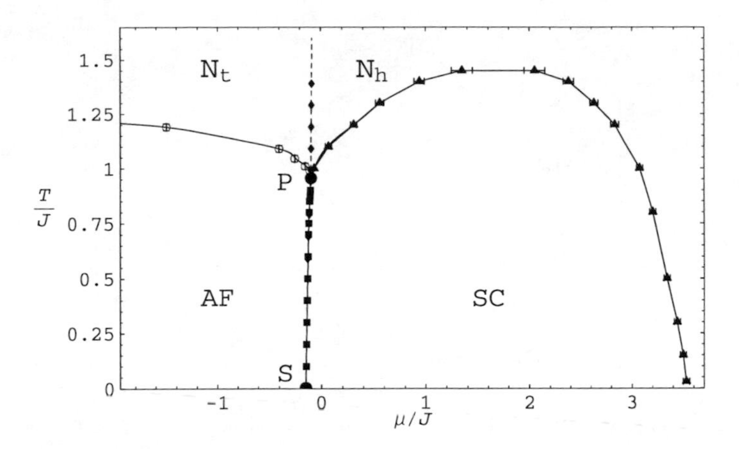

Figure 6.13: Global phase $T(\mu)$ diagram of the three-dimensional pSO(5) model with $J_s = J_c/2$ and $\Delta_s = \Delta_c = J$. N_h is the hole-pair dominated part, N_t the triplet dominated part of the high-temperature phase without long-range order. The separation line between N_h and N_t is the line of equal spatial correlation decay of hole-pairs and bosons.

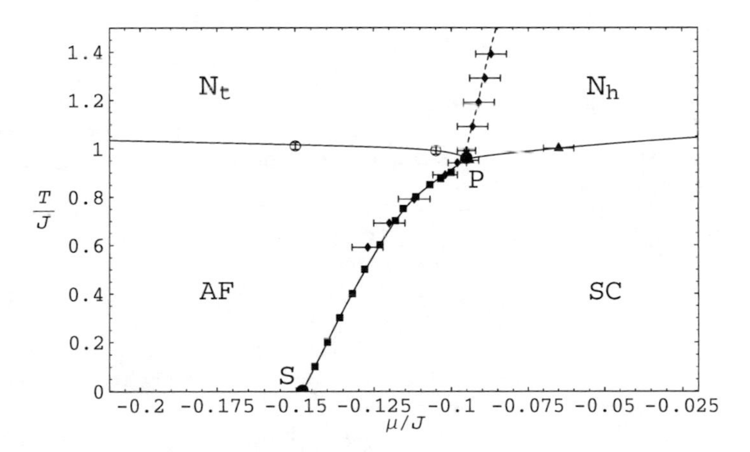

Figure 6.14: Detailed view to the region near the bicritical point in the $T(\mu)$ phase diagram of the 3D pSO(5) model. The filled circles with large error bars trace the line of equal decay behavior of the magnon-magnon and the hole pair-hole pair correlation functions.

phase and an almost hole-free AF ordered phase. Indeed, the results depicted in Fig.

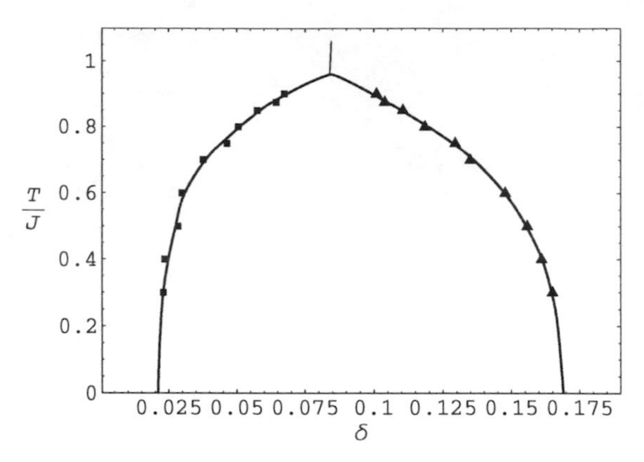

Figure 6.15: Results of a hole-pair density binning analysis along the line from P to S in the previous two figures. The binning shows two distinct peaks in the particle number histogram, whose positions are plotted in this figure. ($\delta = \rho/2$ is the 'physical' hole density as discussed in section 6.1.1.) Hence, the transition line is of first order. The phase separation into a hole-rich and an (almost) hole-free domain ends exactly at the bicritical point, i.e. at $T = 0.96$.

6.15 show a clearly phase-separated regime on the entire transition line from S to P (see Fig. 6.14). When compared to the corresponding 2D results (see Fig. 6.4), the density of the hole-rich phase, $\delta \approx 0.168$ at $T \to 0$, proves to be larger by a factor of about 3/2, whereas the densities of the two hole-depleted phases do not differ significantly.
The impact of phase separation onto the global phase diagram when traced as a function of hole doping δ is shown in Fig. 6.16. As in the HTSC compound La_2CuO_{4+y} (Fig. 6.6), the AF and SC phase are now separated by a "forbidden region" of phase separation.
In section 6.1.2 we have presented a cuprate material which indeed shows this phase separation into AF and SC domains (see 6.5). However, there is a wide range of other SC materials which instead have a mixed phase in which AF and SC order coexist, for example the hybrid ruthenate-cuprate $RuSr_2YCu_2O_8$ [197], the heavy-fermion compound UPd_3Al_3 [198], the classical superconductors [Tm,Er,Ho,Dy]Ni_2B_2C [199] or some organic superconductors like λ-$(BETS)_2Fe_xGa_{1-x}Cl_{4-y}Br_y$ [200].
When discussing the 2D pSO(5) model, we have seen that adding a slight off-site Coulomb repulsion to the pSO(5) model quite effectively destroys the phase separation and forces the system into one single mean-density phase. In 2D we were not able to see whether this mixed phase has mixed AF *and* SC order because the Mermin-Wagner theorem forbids a truly AF-ordered phase in 2D. In 3D, we do not have this problem, hence we can try to show that the mixed phase of mean density of $\delta \approx 0.1$ is indeed both antiferromagnetic and superconducting. To this purpose we restrict the allowed number of hole-pair bosons in our SSE simulations to a narrow range around the desired mean density δ. This amounts to performing simulations in a quasi *canonical* ensemble and not in the *grand canonical* ensemble as usual. The result (Fig. 6.17) is that indeed all states in the formerly forbidden region of phase separation have nonvanishing AF *and* SC order parameters in

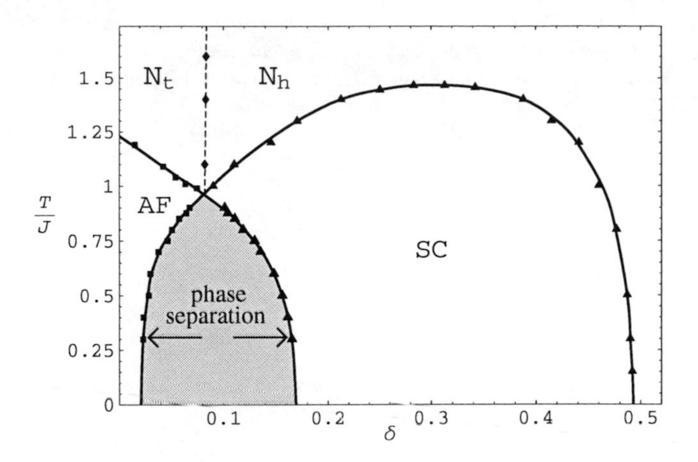

Figure 6.16: Global phase diagram of the 3D pSO(5) model as a function of hole doping $\delta = \rho/2$. The first order transition line from S to P in the $T(\mu)$ diagram becomes a "forbidden region" due to phase separation.

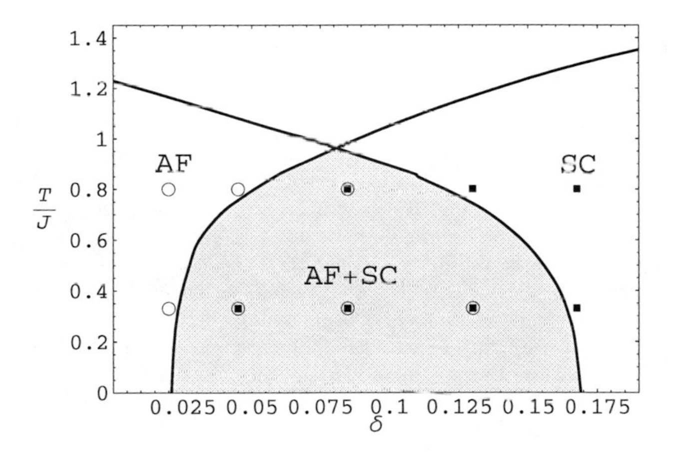

Figure 6.17: The "forbidden region" of phase separation can be accessed by keeping the hole density δ fixed (simulation in the canonical ensemble). (Another way is to introduce off-site Coulomb interactions). The resulting states have both AF and SC long-range order if (and only if) they are situated in the former range of phase separation (shaded region). Open circles symbolize states with AF order, filled squares stand for SC states.

the infinite-volume limit $N_s \to \infty$. This could still be due to large-scale domain separation into hole-rich and (almost) hole-free regions on the lattice. However, the inspection of the spatial correlations between hole-pair bosons shows a 'correlation hole' at distances of one and two lattice spacings, followed by a smooth and continuous decay of the correlation for large distances. This indicates that the charged bosons try to avoid each other as much as possible and are uniformly distributed over the whole lattice.

6.2.4 Spin resonance peak

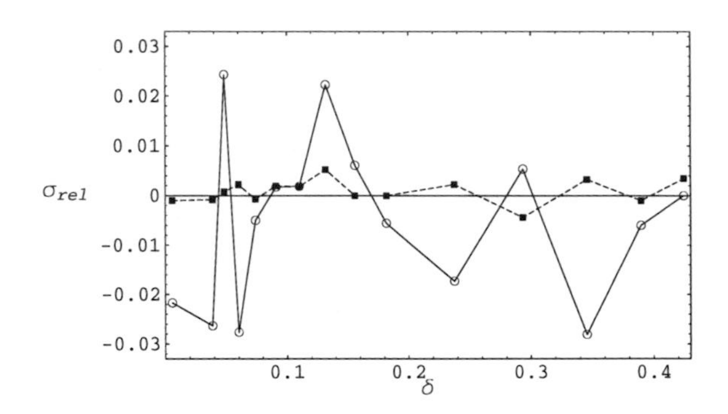

Figure 6.18: Comparison of the spin response at $\mathbf{k}=(\pi,\pi)$ on the $10\cdot10\cdot10$ and the $14\cdot14\cdot14$ lattice. The solid squares connected by a dotted line are the relative differences of the resonance frequencies: $\sigma_{\text{rel}}(\omega) = (\omega_{10} - \omega_{14})/\omega_{14}$. The open circles and solid line are the corresponding differences of the peak weights (i.e. of the areas A under the peak): $\sigma_{\text{rel}}(A) = (A_{10} - A_{14})/A_{14}$.

Next, we are going to study the spin resonance peak of the 3D pSO(5) model at $\mathbf{k}=(\pi,\pi)$. In the preceeding sections, the numerical data shown in the figures were all the result of finite-size scalings based on lattice sizes with up to $16\cdot16\cdot16$ or $18\cdot18\cdot18$ sites. In this section, we restrict ourselves to the $10\cdot10\cdot10$ lattice, which economizes much computer time. In order to justify this procedure, we have once performed a series of calculations in which the spin response of the $10\cdot10\cdot10$ lattice and the $14\cdot14\cdot14$ lattice have been compared. The result is shown in Fig. 6.18. It is clear that there is no systematic deviation between the two lattice sizes, the random differences being of the order of 2% for the peak weights and 0.2% for the resonance frequencies. These differences are most probably entirely due to the non-trivial Maximum Entropy method with which the Green's functions measured by SSE are transformed from imaginary time τ to frequencies ω [20, 107]. We conclude that the $10\cdot10\cdot10$ is sufficiently large to keep finite-size errors below 1 or 2 percent.
The finite-size problem being removed, we proceed to the discussion of the spin resonance data themselves. First we study the resonance frequencies at $\mathbf{k}=(\pi,\pi)$ and the corresponding peak weights as a function of the chemical potential (Fig. 6.19). At low temperatures ($T/J=0.15$ and $T/J=1$), the transition from the AF into the SC phase can easily be identified: at $\mu<-0.1$ the spin response is dominated by a massless spin-wave peak whose weight surmounts the peak weights in all other phases by two orders of magnitude. At the critical μ, the spin wave almost instantaneously vanishes, and the

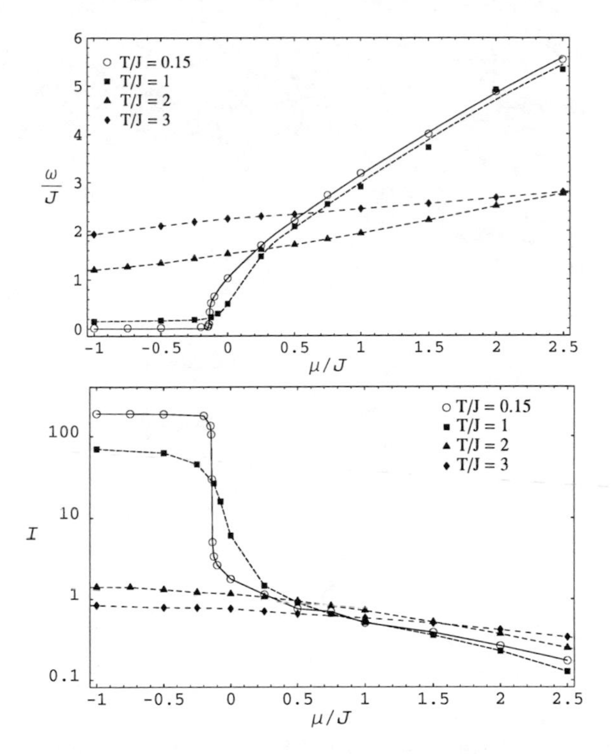

Figure 6.19: Spin resonance frequencies at $\mathbf{k}=(\pi,\pi)$ (top) and the corresponding peak weights (bottom) as a function of the chemical potential for different temperatures T. At $T/J=0.15$ and $T/J=1$ the system crosses the phase transition line between the AF and the SC phase at $\mu/J\approx-0.1$. At higher temperature, the system stays for all μ in a phase without any long-range order.

monotoneously dispersing characteristic peak of the SC phase appears. Just as in the 2D case, the underdoped range is modeled correctly: the resonance frequency increases with T_c, and at optimal doping $\mu_{opt}/J=1.7$ the ratio between $T_c/J\approx1.47$ and the resonance frequency $\omega_{opt}/J\approx5$ is about 0.29, which agrees quite well with the corresponding ratio of 0.20 measured at $YBa_2Cu_3O_{7-\delta}$. At high temperatures $T/J\gtrsim2$, i.e. above T_N and T_c, only a weakly dispersing residual peak appears. What seems surprising, however, is the fact that the weight of the high-temperature peaks surpasses the weight of the "SC peaks" in wide ranges of the hole-pair dominated regime $\mu\gtrsim0$. This seems a contradiction to experiment. A possible explanation might be the fact that for given μ the mean hole-pair density increases with temperature. This bias can be removed if one studies the spin response as a function of density δ (see Fig. 6.20). Now the peak weight in the SC phase is indeed superior to the weights of the high-temperature peaks above the SC phase. However, a sharp decline at the phase transition can not be detected: The peak weight at

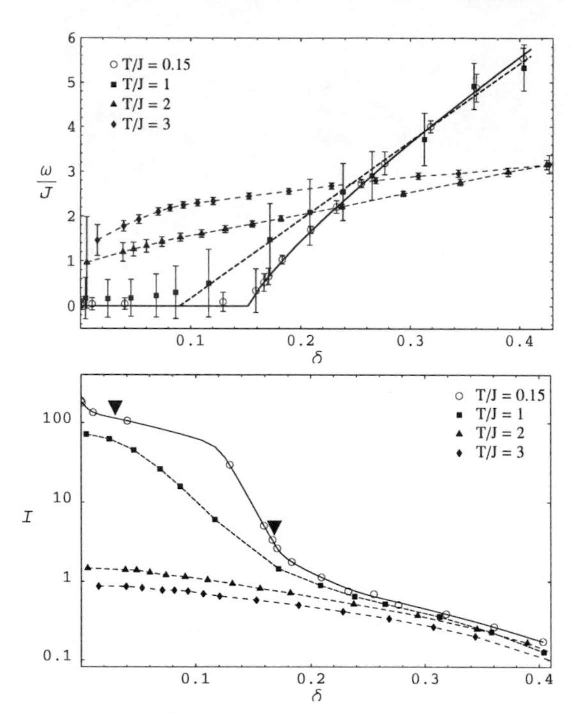

Figure 6.20: Spin resonance frequencies at $\mathbf{k}=(\pi,\pi)$ (top) and the corresponding peak weights (bottom) as a function of the hole-pair density δ for different temperatures T. At $T/J=0.15$ and $T/J=1$ the decline of the spin wave peak within the regime of phase separation (indicated by the two black triangles) is clearly visible.

$T/J=3$ is still about half the weight at $T/J=0.15$. Probably most of the weight at high temperatures is due to the magnons in the systems, whose density increases for given δ with temperature.

6.2.5 Restoration of SO(5) symmetry at the bicritical point

In this section we perform a scaling analysis similar to the one performed by Hu [30] in a classical SO(5) system. The most interesting result of this analysis will be the strong numerical evidence that in a region around the bicritical point ($T_b=0.96$, $\mu=-0.098$) the full SO(5)-symmetry is approximately restored, a very non-trivial property for a system whose SO(5)-symmetry has manifestly been broken by projecting out all doubly-occupied states.

First we want to determine the form of the $T_N(\mu)$ and $T_c(\mu)$ curves in the vicinity of the bicritical point. An important property of SO(5)-symmetry is that the two curves should merge tangentially into the first order line. However, the two equations (6.9) in Hu's classical approach are only applicable if the first order transition line is exactly *vertical*

in the $T(\mu)$ phase diagram, which is not the case in the 3D pSO(5) model (see Fig. 6.14). This means that μ is not a scaling variable for the bicritical point.
A solution to this problem is to perform a transformation from the old μ axis to a new μ' axis defined by

$$\mu'(T) = \mu - (T - T_b)/m\,, \tag{6.11}$$

where m is the slant of the first order line below T_b. Figure shows the resulting $T(\mu)$ phase

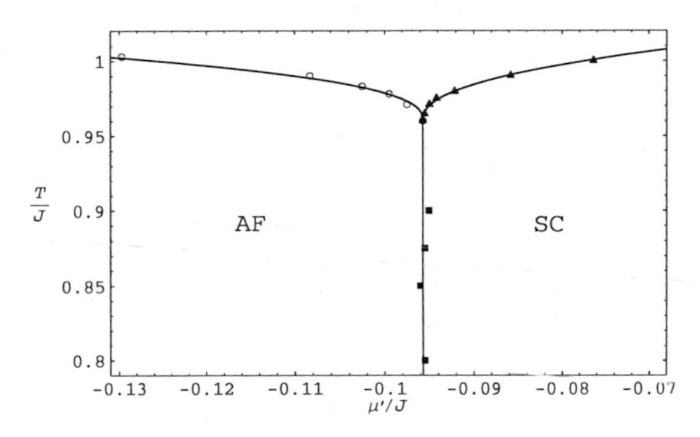

Figure 6.21: $T(\mu)$ phase diagram of the pSO(5) model after performing the transformation $\mu' = \mu - 0.11(T - T_b)$ (with T_B=0.96 J). This tilts the μ values of each point with the bicritical point (μ_b; T_b) remaining fixed; the effect is that the first order AF-SC transition line's deviation from the vertical axis is removed. This makes the vicinity of the bicritical point comparable to the corresponding region in X. Hu's classical SO(5) model.

diagram in the vicinity of the bicritical point after the transformation with m=0.11. The similarity to X. Hu's classical phase diagram is striking, and the two fitting functions (6.9) describe the transformed $T_N(\mu')$ and $T_c(\mu')$ curves quite well. This fit is shown in more detail in Fig. 6.23 The figure shows that both fits succeed very well, yielding

$$\begin{aligned} B_2 &= 0.220 \pm 0.02 \\ B_3 &= 0.165 \pm 0.012 \\ \phi &= 2.35 \pm 0.5 \end{aligned}$$

Within error bar accuracy, the constants B_2 and B_3 indeed satisfy the relation $B_2/B_3 = 3/2$; ϕ, on the contrary, is considerably larger than 1.387, the value determined by Hu. However, it should be noted that above determination of ϕ is not very accurate: the data points in Fig. 6.22 are the result of a delicate finite-size scaling, followed by the transformation from μ to μ' which again increases the numerical error bars. For this reason it cannot be excluded that the difference in the ϕ values is mainly due to statistical and finite-size scaling errors.
In Fig. 6.23 we show how the data points for Fig. 6.22 were generated by a finite-size scaling with lattice sizes up to $18 \cdot 18 \cdot 18$. On the SC side, the finite-size scaling turns out to be quite reliable, on the AF side, the fluctuations in the particle numbers of the

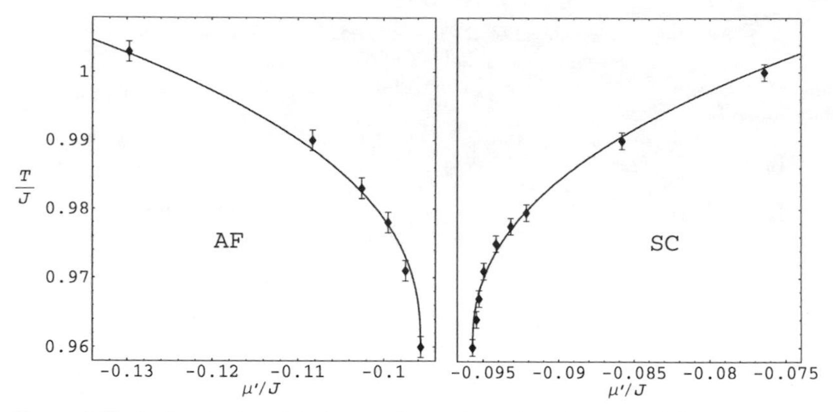

Figure 6.22: In the vicinity of the bicritical point ($\mu_b; T_b$) in the tilted $T(\mu')$ phase diagram, the Néel temperature can be described by $T_N(\mu')/T_b - 1 = B_3(\mu_b - \mu')^{1/\phi}$ with $B_3 = 0.165 \pm 0.015$ and $1/\phi = 0.41 \pm 0.04$ (left). The critical temperature is fitted by $T_c(\mu')/T_b - 1 = B_2(\mu' - \mu_b)^{1/\phi}$ with $B_2 = 0.22 \pm 0.015$ and $1/\phi = 0.43 \pm 0.04$ (right).

three triplet bosons slightly increase the statistical errors of the SSE results and make the finite-size scaling more difficult.

The critical exponents for the onset of AF and SC order as a function of temperature for various chemical potentials can be determined from Fig. 6.24. Far into the SC range, at $\mu = 1.5$, we find

$$\Upsilon \propto (1 - T/T_c)^{\nu} \quad \text{with} \quad \nu = 0.66 \pm 0.02\,,$$

which matches very well the value of 0.666 ± 0.004 derived by Hu and agrees with RG results for the pure 3D XY model. When the bicritical point is approached, the value of ν increases to

$$\begin{aligned} \nu(\mu=0.2) &= 0.77 \pm 0.03 \\ \nu(\mu=0.003) &= 0.89 \pm 0.06. \end{aligned}$$

This is in accordance with Hu's observations, too. In both models, the deviation from the XY exponent $\nu = 0.666$ seems to be due to the fact that the criticality region shrinks considerably when μ' approaches the bicritical value μ_b: since the numerical determination of ν requires data points from a finite temperature range, a shrunken criticality region inevitably yields biased results for ν if it causes some of the used data points to drop out of this region .

On the AF side, error bars are larger. From Hu's formula $\sqrt{C_{AF}} \propto \left(1 - T/T_N(\mu)\right)^{\beta_3}$ we obtain

$$\begin{aligned} \beta_3(\mu=-0.15) &= 0.385 \pm 0.06 \\ \beta_3(\mu=-0.4) &= 0.325 \pm 0.05 \\ \beta_3(\mu=-2.25) &= 0.35 \pm 0.04 \end{aligned}$$

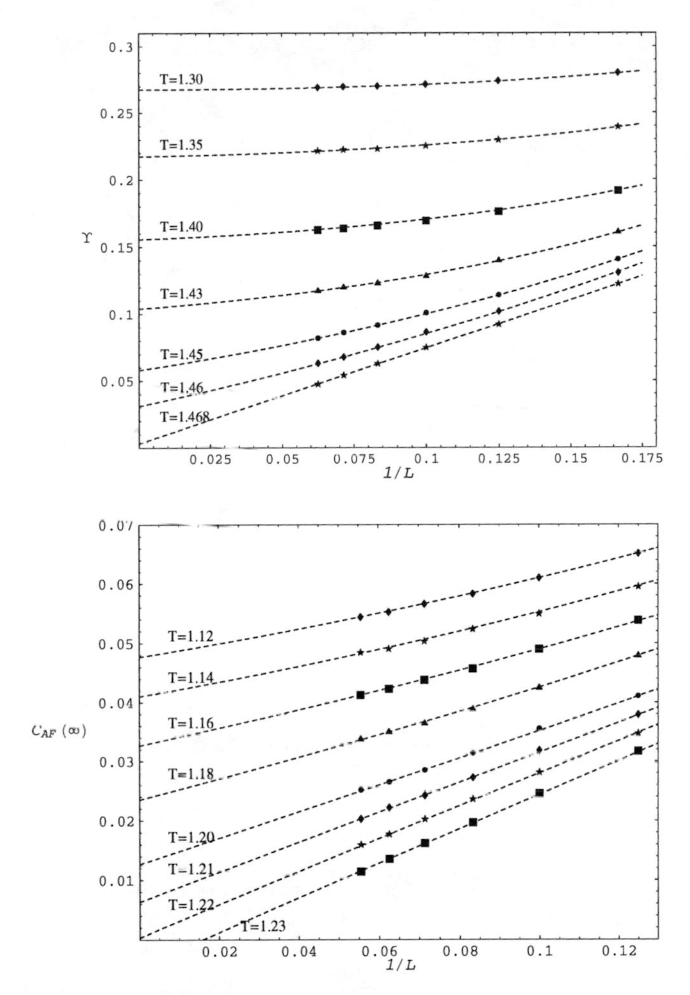

Figure 6.23: The data points in Fig. 6.22 have been obtained from a finite-size scaling with lattices up to $18\cdot 18\cdot 18$. On the SC side of the phase diagram we trace the helicity $\Upsilon(T)$ at $\mu/J = 1.5$ (top). On the AF side, long-range antiferromagnetic order corresponds to a non-vanishing value of the squared AF order parameter $C_{AF} = \frac{1}{N_s^2}\sum_{\alpha,i,j}(n_\alpha(i)n_\alpha(j))$ on the infinite lattice, with i and j running over all lattice sites and $n_\alpha(i)$ being the three components of the Néel order parameter on site i (bottom); this plot has been recorded at $\mu/J = -2.25$.

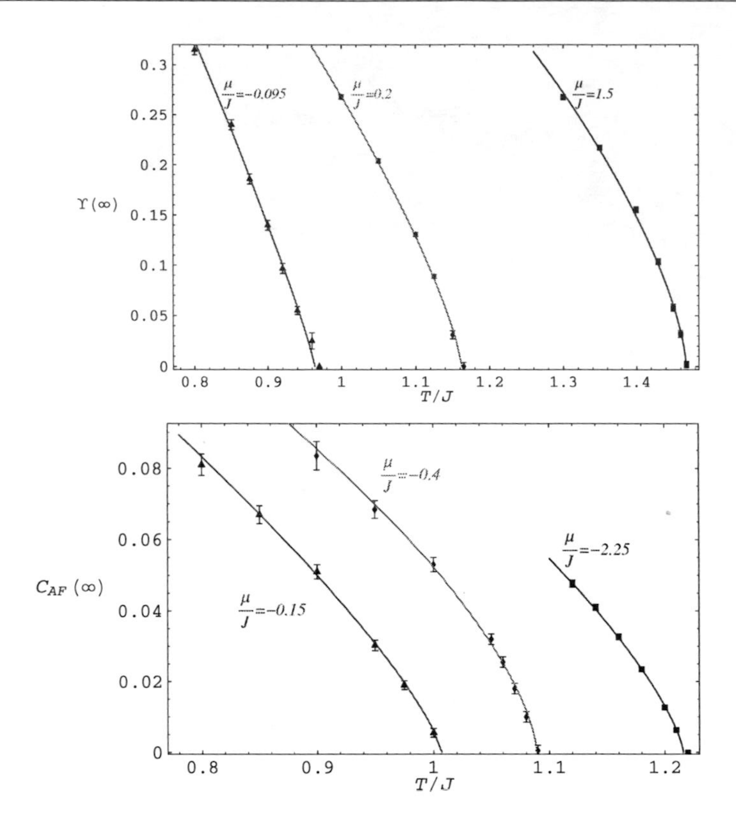

Figure 6.24: The superconducting (top) and antiferromagnetic (bottom) order parameters as functions of temperature for various chemical potentials μ.

This again is in good accordance with the value of $\beta_3 = 0.363$ given by Hu for $\mu = -0.5$, and it also agrees with RG results.
After it turns out that from Fig. 6.22 an exact determination of ϕ is impossible we try the more promising approach of using formulas (6.7) and (6.8). The result is shown in Fig. 6.25: we determine the ratio

$$\nu_5/\phi = 0.52 \pm 0.01,$$

which is in excellent accordance with Hu's reference value of 0.525 ± 0.002.
The determination of ϕ itself then succeeds using (6.8). In Fig. 6.26 we have applied (6.8) onto 9 different combinations of $(\mu_1, \mu_1' = \mu_2, \mu_2')$ values with $\mu_1/\mu_1' = \mu_2/\mu_2' = 0.5$. The result is

$$\phi = 1.43 \pm 0.05\,,$$

which is in good agreement with X. Hu's reference value of 1.387.

Altogether, the scaling analysis of the 3D pSO(5) model has produced critical exponents and crossover parameters which match well with the corresponding values obtained from a classical SO(5) model. This gives very convincing indications for the starting hypothesis of this section: that at the bicritical point of the 3D pSO(5) model the full SO(5) symmetry is approximately restored. Specifically, either the bicritical point lies in the domain of attraction of the O(5) Heisenberg fixed point, and the latter is stable, or this fixed point controls the scaling behavior in a large, numerically relevant, region around the bicritical point. It is difficult, in fact, to conclude whether the O(5) fixed point controls the ultimate scaling, and, in fact, strong analytical statements seem to indicate that this is not the case [180]. Nevertheless, it is remarkable that a system like the pSO(5) model, which is so far from SO(5) symmetry, is driven by scaling such close to an exact SO(5)-invariant fixed point.

6.3 Four-boson models beyond the pSO(5) Model

The pSO(5) model proved capable to reproduce a lot of features of real cuprates. However, for the settings $J_s = J_c/2$ and $\Delta_s = \Delta_c = J_s$ the ratio T_c/T_N adopts unrealistic values of more than 1. This problem can be surpassed if the fixed ratio $J_s = J_c/2$ is abandoned, which amounts to starting from a model whose ground state is not SO(5)-symmetric. Figure 6.27 demonstrates that by varying J_s/J_c the critical temperature can be decreased to realistic values. The value of T_N is not affected by this change.
This result demonstrates once more that four-boson models are very promising model systems for the description of high-T_c superconductors

6.4 summary

In summary, we have shown that the projected SO(5) model in two and three dimensions – or more general four boson models of type (1.15) – can be considered as a generic model for many novel superconducting materials. It gives a semiquantitative or even quantitative description of many properties of the HTSC in a consistent way. In particular, we have identified an AF and a SC phase whose bounds look similar to the the ones of real cuprate materials. Also, the doping dependence of the chemical potential as well as that of the neutron resonance peak in the underdoped regime are reproduced correctly. Furthermore, the pSO(5) model reproduces effects like phase separation or coexistence of antiferromagnetism and superconductivity.
A detailed scaling analysis of the 3D pSO(5) model has produced critical exponents and crossover parameters which match surprisingly well with the corresponding values obtained from a classical SO(5) model. This gives strong evidence that the bicritical point of the 3D pSO(5) model is controlled, at least in a large scaling regime, by a completely SO(5)-symmetric fixed point.

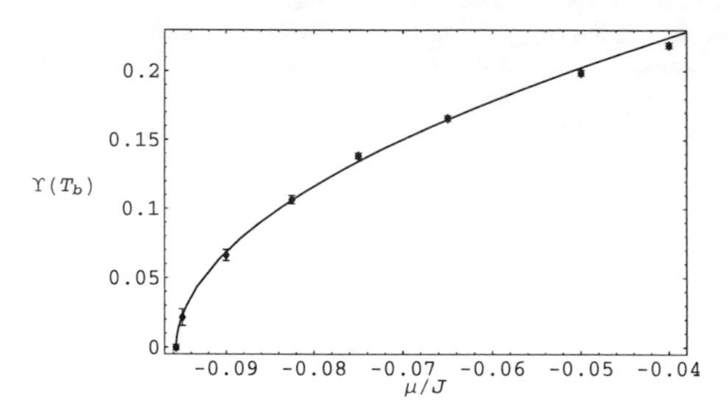

Figure 6.25: Helicity Υ as a function of the chemical potential μ at $T=T_b$. From this function, the value of ν_5/ϕ can be obtained via equation (6.7). The result is: $\nu_5/\phi=0.52\pm0.01$.

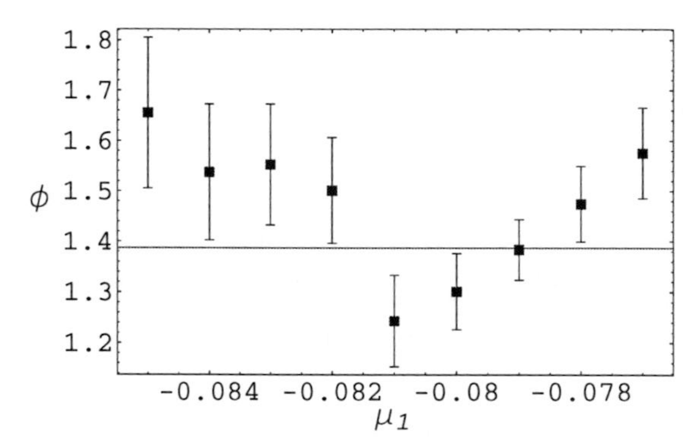

Figure 6.26: Determination of ϕ via formula (6.8). We have calculated $\Upsilon(T,\mu)$ at $T=T_b$ and $T=T_b\pm0.01$ for 9 different combinations $(\mu_1,\mu_1'=\mu_2,\mu_2')$ with $\mu_1/\mu_1'=\mu_2/\mu_2'=0.5$ from the range $0.85\le\mu_1\le-0.77$. The horizontal line is the ϕ value obtained by X. Hu in the classical SO(5) model.

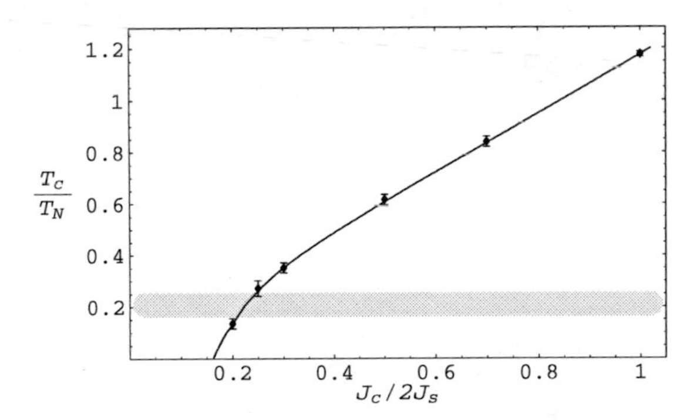

Figure 6.27: Ratio $T_c : T_N$ as a function of $J_c : 2J_s$: the SO(5)-symmetric combination $J_c\!=\!2J_s$ yields an unrealistically large ratio of $T_c : T_N\!=\!1.18$. Realistic values in the cuprates (see Fig. 5.2) are 0.17 to 0.25 (shaded band). This ratio can be reached by decreasing J_c to about $J_s/2$.

Summary

This thesis was divided into two major parts. In the first, 'technical' part, an overview on currently available computational techniques for correlated electron systems was given as were own contributions to the further development of three of these techniques, Exact Diagonalization (ED), the Quantum Monte Carlo technique of Stochastic Series Expansion (SSE) and computer aided algebraic manipulations (CAAM). Several program codes were described and used to get new insights into various current research areas of theoretical solid state physics. In the second, 'physical' part, new contributions to the author's main research area, the projected SO(5) theory of high temperature superconductivity, were given.

The first part began with an overview on frequently studied microscopic models for many-electron systems and on important numerical simulation techniques. In addition, we showed how microscopic fermionic models describing two-dimensional electron systems such as the CuO_2 planes in the cuprates can be mapped onto 'coarse-grained' effective bosonic models. In particular, the low-energy spin and charge excitations of the t-J and Heisenberg model are correctly captured by a four-boson model in which three hard-core bosonic quasiparticles describe triplet excitations and one charged boson models the formation of hole pairs.

Chapter 2 was dedicated to ED and its applications. It started with a discussion of strengths and limitations of this method and an overview on successful approaches for an efficient implementation on modern computer architectures. Two main application fields, which need different coding concepts and implementation tricks, were identified and described: the simulation of low-dimensional regular lattices with periodic boundary conditions, and of single unit cells or metal-ligand clusters with a complex internal structure. At the end of the chapter, two applications of ED on current research topics were given. First, an ED study on metal-insulator transitions in the half-filled Hubbard model with second-nearest neighbor hopping at zero temperature revealed a rich phase diagram. Two different metallic phases were identified and two types of insulating phases: an AF ordered Mott-Heisenberg insulator and a Mott-Hubbard insulator, in which the insulating behavior is uniquely due to electrostatic electron-electron interaction. The Hubbard model is a widely accepted simple effective model for the CuO_2 planes in cuprate superconductors. In a second exemplary application of ED methods, experimentally recorded

X-ray magnetic circular dichroism spectra (XMCD) of the technologically interesting material CrO_2 were reproduced within ED simulations of a single CrO_6 cell. This mapping of real experiment and "computer experiment", which succeeded surprisingly well, can help to understand the most important elementary physical processes in this material. Furthermore, it permits to determine the values of microscopic interaction parameters such as crystal fields, Coulomb and exchange integrals, or spin-orbit couplings.

In Chapter 3, Stochastic Series Expansion (SSE) was reviewed and a significant extension of this QMC technique was developed: a highly efficient method to measure arbitrary Green's functions during the construction of SSE's loop updates. In benchmark tests on Heisenberg models in one, two and three dimensions we demonstrated that SSE's scaling behavior of computational effort as a function of system size and inverse temperature rivals the excellent scaling properties of the loop algorithm. SSE clearly outperforms the loop algorithm if external fields or chemical potentials are to be treated. In addition, SSE is relatively easily applicable to wide classes of Hamiltonians. The benchmark tests also proved the efficiency of the new mechanism of measuring Green's functions. From the example of the $\langle S^z(\mathbf{r},\tau)S^z(\mathbf{r}',\tau')\rangle$ and $\langle S^+(\mathbf{r},\tau)S^-(\mathbf{r}',\tau')\rangle$ measurements in the Heisenberg model we concluded that measuring highly nontrivial Green's functions can even be the preferred method of determining also the diagonal real space dynamical correlation functions. The SSE codes created during this thesis were then applied to a wide range of problems arising from current solid state physics. This demonstrates both the performance of the SSE technique and the flexibility of the chosen software design model, in which we tried to separate the SSE core algorithms, the lattice geometries and the physical model (particles and their interactions) into independent object oriented software modules. These software modules interact via well-defined interfaces, so that within a few seconds a new application of SSE on a certain lattice geometry and a certain physical model can be composed. New results gained from three exemplary applications of SSE were sketched at the end of chapter 3: a study on the destruction of superfluid and long range order by impurities in two-dimensional systems, a work on quantum phase transitions in the two-dimensional hardcore boson model, and a bosonic description of the spin-1/2 Heisenberg ladder with nonmagnetic impurities, a model system for zinc or nickel doping in cuprate superconductors.

We ended the 'technical' part of this thesis with some notes on how computers can be used to ease arduous algebraic term transformations arising from quantum mechanical many-particle problems. In a second part of the chapter the Mathematica^{TM} program developed from these ideas was used for an analytical study of the Hubbard model. We have investigated the most important corrections over the Hubbard-I approximation in the limit $U/t \to \infty$ and electron density $n = 1$. The Hubbard-I approximation turned out to describe charge fluctuations on a 'background' of singly occupied sites. The charge fluctuations are point-like, and correspond to an electron moving between empty sites and an electron moving between singly occupied sites. In our extended scheme for the Hubbard model we have augmented these point-like charge fluctuations by additional 'particles' which are composite in character and consist of a point-like charge fluctuation together with a spin-, density-, or η-excitation. Comparison of the obtained single-particle spectral density with QMC results for a variety of systems showed a quite reasonable agreement. In particular the apparent 4-band structure seen in the numerical spectra finds its natural explanation in the extended Hubbard approximation. We also note

that QMC simulations where the spectra of the composite excitations have actually been computed, further support the present interpretation. We thus have a quite successful method of computing the full quasiparticle band structure of the Hubbard model, at least in the paramagnetic case and at half-filling.
The second main part of this work was dedicated to one larger physical study: to a numerical analysis of the projected SO(5)-symmetric model of high-temperature superconductivity. This model maps the low-energy physics of the CuO_2 planes in the cuprate superconductors onto a model consisting of 4 bosonic quasiparticles: three triplet excitations and a hole-pair excitation. We started the study with a short review on the basic principles and ideas of SO(5) theory and projected SO(5) theory. Then we demonstrated that the projected SO(5) model – which is the simplest bosonic model containing two key ingredients for the HTSC: the Mott gap and the vicinity of AF and SC phases – gives a semiquantitative description of many properties of the HTSC in a consistent way. In particular, an AF and a SC phase were identified whose bounds look similar to the ones of real cuprate materials. Also, the doping dependence of the chemical potential as well as that of the neutron resonance peak in the underdoped regime are reproduced correctly. Furthermore, the pSO(5) model was shown to reproduces effects like phase separation or AF-SC coexistence which appear in some novel superconducting compounds. Finally, a scaling analysis of the 3D pSO(5) model produces critical exponents and crossover parameters which match well with the corresponding values obtained from a classical SO(5) model. This gives strong evidence that the bicritical point of the 3D pSO(5) model is controlled, at least in a large scaling regime, by a completely SO(5)-symmetric fixed point.
Detailed appendices to this work provide documentations for the program packages and algorithms developed during this thesis: the two different Exact Diagonalization codes, the computer algebra framework for quantum mechanical term manipulations, the SSE program package, and the new approach to 'selective' Fourier transformations.

Program package documentations

This appendix describes organization, features and usage of four program packages which implement the ideas and concepts discussed in the first three chapters of this work.

- ed_hubb an ED program for low-dimensional regular lattices with periodic boundary conditions and for Hubbard-like physical systems with four allowed states pers site. Supported energetic interactions are single-particle hopping, Coulomb interaction and Heienberg spin-spin-exchange on nearest-neighbor and second-nearest neighbor sites.

- Program ed_clust is an ED implementation written in C++ which models one single unit cell of a metal-nonmetal compound with one central p-, d- or f-metal atom and an arbitrary number of nonmetal ligand orbitals.

- Program sse casts the method of stochastic series expansion as described in chapter 3 into a C++ code package that works for almost arbitrary Hamiltonians and lattice geometries.

- Program QM provides an easily extensible Mathematica© framework for algebraic manipulations of quantum mechanical expressions in the formalism of Second Quantization.

A.1 Program package "ed_hubb"

A.1.1 Program files and documentation

Program package ed_hubb contains the following files:

1. main program files:

`ed_main.cc`	main file for project 'ed_hubb'
`ed_H_ket.cc`	Hamiltonian class and operator application $H\cdot\|x\rangle$
`ed_base.cc`	Hilbert space basis classes
`ed_c_k.cc`	operators $c(\mathbf{k})$ and $c^\dagger(\mathbf{k})$
`ed_n_k.cc`	operators $n(\mathbf{k})$ and $S^z(\mathbf{k})$
`ed_comps.cc`	analyses states; finds and displays the most important components
`ed_corr.cc`	static n-n and S^z-S^z correlations
`ed_input.cc`	interprets the parameter input file
`ed_lanc.cc`	Lanczos algorithm for calculating energy eigenstates and dynamical response spectra
`ed_mixed.cc`	small auxiliary routines
`ed_spin.cc`	calculates total spin and z-spin expectation values

2. auxiliary plotting package:

`edplot.cc`	main file for plotting program 'edplot'
`ed_lrntz.h`	classes for replacing δ-peaks by Lorentzians
`ed_lrntz.cc`	implementation file for `ed_lrntz.h`
`plotpnts.h`	determine plot points: higher point density in regions around peaks
`edplot_input.eps`	format and frame settings for postscript file to be generated

3. auxiliary general purpose header files:

`mymath.h`	mathematical functions
`mymath.cc`	implementation file for '`mymath.h`'
`myvalar.h`	array classes
`myvalar.ih`	template implementations for '`myvalar.h`'
`myvalar.cc`	non-template implementations for '`myvalar.h`'

4. documentation, parameters and settings:

`Makefile`	makefile for creating programs '`edmain`' and '`edplot`'
`ed_defs.h`	compiler and linker settings, program mode flags
`input.txt`	parameter file for '`edmain`'
`pltparam.txt`	parameter file for '`edplot`'
`sx4notes.txt`	hints for using NEC SX4/SX5 vector computers

A.1.2 Usage hints

The two default parameter files `input.txt` and `pltparam.txt` contain comments lines and should be self-explaining. Just replace the default parameters by the desired settings. Program `edmain` works on 1D chains, two-leg ladders and 2D rectangles; it can deal with Hamiltonians consisting of on-site and off-site Coulomb interaction (U, V), single-particle hopping (t) and Heisenberg spin-spin interaction (J). For the off-site interaction constant ((V, t and J) there are two settings in the parameter file, e.g. `V=` and `V2=`. In 1D chains and 2D rectangles, the first setting applies for nearest neighbor bonds, the second setting applies for second-nearest neighbor bonds. On the two-leg chain, the first setting defines the interaction strength on nearest neighbor bonds in ladder direction, the second one on bonds along the rungs. The sign convention for all interaction constants is that positive values increase the system's energy, negative values decrease it. (Thus, one must set $t<0$ to define a single-particle hopping which decreases the systems energy.).

Program `edmain` produces a protocol output file `prot????.txt` and several dynamical response output files: `pes_??` for photoemission `ipes??` for inverse photoemission, `deco??` for dynamical density correlations and `spco??` for spin correlation functions. The numbers appearing at the place of `??` run over all allowed momenta. The dynamical response data files consist of two columns of ASCII data separated by blanks. The first row contains the momentum transfers k_x and k_y, the second row the ground state energy and the norm of the resulting state after operator application ($c(\mathbf{k})$ for photoemission, $c^\dagger(\mathbf{k})$ for inverse photoemission, $n(\mathbf{k})$ for density correlation and $S^z(\mathbf{k})$ for spin correlation spectra). The third row contains the number n of δ-peaks recorded by the Lanczos algorithm; the energies and peak weights of these n poles are given in the following n rows of the data files. The last two rows, finally, record the number of spin-up and spin-down electrons and the expectation values of total spin and of z-spin.

A.2 Program package "ed_clust"

A.2.1 Program files and documentation

Program package `ed_clust` consists of 3 groups of files:

1. main program files:

`ed_main.cc`	main file for project '`ed_clust`'
`ed_hbase.h`	Hilbert space basis classes
`ed_hbase.ih`	template implementations for '`ed_hbase.h`'
`ed_hbase_.cc`	non-template implementations for '`ed_hbase.h`'
`ed_H_ket.h`	Hamiltonian class and operator application $H\cdot\|x\rangle$
`ed_H_ket.ih`	implementations for '`ed_H_ket.h`'
`ed_op.h`	angular momentum operators: $\boldsymbol{S}$, $\boldsymbol{L}$, $\boldsymbol{J}$, S_z, L_z, J_z, and X-ray absorption operator
`ed_op.ih`	template implementations for '`ed_op.h`'
`ed_op_.cc`	non-template implementations for '`ed_op.h`'
`ed_lanc.h`	Lanczos algorithm for calculating energy eigenstates and dynamical response spectra
`ed_lanc.ih`	template implementations for '`ed_lanc.h`'
`ed_spectrum.h`	classes for combining different dynamical response spectra and for Lorentz-smoothing
`ed_spectrum_.cc`	implementations for '`ed_spectrum.h`'

2. auxiliary general purpose header files:

`array.h`	header file implementing array and vector classes
`array.ih`	template implementations for 'array.h'
`array_.cc`	non-template implementations for 'array.h'
`matrix.h`	header file implementing matrix classes, matrix-matrix and matrix-vector operations
`matrix.ih`	template implementations for 'matrix.h'
`my_sort.h`	some sorting routines needed in Hilbert basis classes
`read_param.h`	a class for reading parameter files and extracting parameter values

3. Documentation, parameters and settings:

`tar_ed_clust`	shell script writing all files of project `ed_clust` into an archive file `ed_clust.tar.gz`
`ed_param.txt`	an example parameter file with explaining comments
`debug_protocol.h`	debug tools: macros and functions for protocol writing, range checking and memory allocation checks.
`Makefile`	makefile for creating program 'edmain' by typing 'make edmain'
`make.inc`	auxiliary file for 'Makefile': compiler, linker and library paths. This file probably has to be modified when transfering the `ed_clust` package to a new computer system

A.2.2 Parameters

Program `edmain` needs a parameter file containing information on the system to be simulated and the quantities to be calculated. The syntax for starting `edmain` is "`edmain` *param_file*"; calling "`edmain -h` " or "`edmain ?` " makes `edmain` print some usage hints and a list of recognized parameters in the parameter file. Such a file should be a simple unformatted text file containing a series of settings of the form "*param_name* = *value*" with each setting on a separate line. Lines beginning with '//' are considered comments and ignored.
One parameter file can be used to start several program runs with different parameter values. For this purpose one can assign a series of values (separated by commas) to some or all of the parameters. The total number of runs to be performed is determined by the parameter value series with maximum length N: run n is assigned the n-th value in this series. For other parameters there might be less than N values on the right hand side of the setting equation. These shorter series of values are traversed more than once: if the last value in the series has been reached and if there are further program runs to be started, then `edmain` jumps back to the start of the series and in the following run the corresponding parameter is again assigned the first value of the series.
The parameters recognized by `edmain` can be grouped into 7 subsets:

- Protocol writing and checks to be performed
- Fine-tuning of the Lanczos method
- Automatic result file naming

- Number of orbitals and electrons
- Orientation of unit cell and magnetic field
- Energetic interaction constants
- Program mode; which spectra are to be calculated

Each of these subsets will be described in more detail in the following subsections.

Protocol writing and checks to be performed

calc_sum_rules :

If this flag is activated several sum rules concerning X-ray absorption operators for left-circular, right-circular and linear-polarized X-rays are tested for correctness. Allowed values: 0 (= off) and 1 (= on). Default is 0.

H_file
H_file_format
H_char :

By assigning a file name to 'H_file' one can make edmain write all non-zero elements of the Hamiltonian matrix into that file, with 'H_char' being the character (or character string) to be inserted as the name of the Hamiltonian matrix. 'H_file_format' defines the output format:
H_file_format=fortran produces FORTRAN output (indices starting with 1, e.g. H(1,3)=7)
H_file_format=mathematica produces Mathematica© output (indices starting with 1, e.g. H[[1,3]]=7;)
H_file_format=free produces a free format inlcuding a graphical representation of the bra and ket state (indices starting with 0, e.g. H(0,2)=<ovooo|H|voooo>=7).

LSJ_file :

If a file name has been assigned to this parameter, then edmain writes the matrix elements of the angular momentum operators L^2, S^2 and $J^2 = (\boldsymbol{L} + \boldsymbol{S})^2$ into this file.

write_H_auxiliaries :

If activated (i.e. being assigned a non-zero value) this flag makes edmain print detailed information on the construction of the Hamiltonian matrix: what are the energetic interactions, and which non-zero matrix element is due to which elementary interaction.

write_ground_state
write_all_states :

Write the ground state into the protocol file? Or write ground state and all starting states for calculations of spectral functions into the protocol file? Allowed values are 0 (= off) and 1 (= on). Default is 0.

Fine-tuning of the Lanczos method

`peak_thresh` :

δ-peaks in dynamical response spectra with peak weights below this threshold are neglected. Allowed values are real numbers larger than 10^{-7}. Default is 10^{-5}.

`spectra_error` :

Convergence criterion for calculating dynamical response spectra: let $f_n(\omega)$ be the spectral function after n Lanczos steps; the iteration process is aborted if the integrated difference $\int |f_n(\omega) - f_{n+1}(\omega)|$ is smaller than `spectra_error` times $\int f_n(\omega)$.

`max_lanc_it` :

Abort criterium for Lanczos process: if after `max_lanc_it` iterations no convergence has been obtained (in ground state or spectra calculations) then the process is aborted. Default value is 200.

Automatic result file naming

`spectra_file` :

Assigning parameter `spectra_file` a name is the first possibility to define names for spectra output file and protocol output file (the latter has the additional suffix '`.prot`'.

`output_naming_parameters`
`output_naming_texts` :

Leaving `spectra_file` empty and activating automatic output file naming by assigning one or more parameters (separated by ':') to `output_naming_parameters` is the second possibility to define names for output files; for each parameter name appearing in `output_naming_parameters` a corresponding text must be given in `output_naming_texts`. These texts, followed by the corresponding parameters' values, then form the name of the output file.
Example: if one wants to perform two runs, the first one with `nEl=2,B_field=0`, the second one with `nEl=2,B_field=0.05`, then one might define '`output_naming_parameters=nEl:B_field`' and '`output_naming_texts=myrun_nEl:_B`', which produces the output files `myrun_nEl2_B0` and `myrun_nEl2_B0.05`.

Number of orbitals and electrons

`nEl` :

total number of electrons. Default is 1.

`nMetalOrb` :

Number of valence orbitals of the central metal atom. Must be 3 for p-metals, 5 for d-metals, 7 for f-metals. Default is 5.

`nLigandOrb` :

Number of ligand orbitals. Default is 0.

`nCoreStates` :

Number of different states of the metal's core electrons. If there is no core-hole, `nCoreStates` is 1. If by an Xray absorption process a core electron has been transfered into one of the metals valence orbitals, the number of core states is $2\cdot(2\cdot(L-1)+1)$ with L being the orbital angular momentum of the metal's valence shell (i.e. $L=1$ for p-metals, $L=2$ for d-, $L=3$ for f-).

`nDiffMetalOccup` :

For realistic values of the interaction constants the cluster's lowest-energy configurations have only the minimum possible number m of electrons in the metal atom's valence orbitals; the more electrons hopp from ligand orbitals to metal orbitals, the higher the configuration's energy and the lower its contribution to the ground state of the system. In many cases, about 99.9% of the total weight of the ground state are contributed by contributions with m, $m+1$ and $m+2$ electrons on metal orbitals. In this situation one can work in a restricted Hilbert space in which all states with more than $m+2$ or $m+3$ electrons on metal orbitals are projected out. This is done by setting '`nDiffMetalOccup=3`' or '`nDiffMetalOccup=4`'. Default value for this parameter is 100.

Unit cell orientation with respect to the magnetic field

`theta_ax` :

Angle between z-axis (and magnetic field $\boldsymbol{B}\,\|z$) and the unit cell's main axis. Default vaue is 0.

`ph_ax` :

Rotation angle beween the unit cell and the x-axis (which is defined by the direction of the X-ray beam). Default value is $\pi/2$.

Energetic interaction constants

`B_field` :

External magnetic field (parallel to z-axis).

`J_spin_orb_C`
`J_spin_orb_M` :

Spin-orbit coupling on core states and on metal orbitals.

`Delta` :

Orbital energy difference $E_{(ligands)} - E_{(metals)}$ between metal and ligand orbitals.

`orb_energy` :

Orbital energy fine-tuning. Type each non-zero offset in format '*index,value*', separate adjacent offsets by ';' or '|'. (Indices `0..nLigandOrb-1` are ligand orbitals, `nLigandOrb..nLigandOrb+nMetalOrb-1` are metal orbitals.)
Example: `orb_energy = (1:-1.8; 4:-7)`.

`t_ML` :

Hopping elements between metal and ligand orbitals. Type each non-zero element in the form '*met,lig:value*' (with *met* and *lig* being orbital numbers); separate adjacent matrix elements by ';' or '|'.
Example: `t_ML = (0,0:1.5 | 1,0:0.5 | 2,3:-0.2)`.

`V_cf` :

Crystal field potentials between two metal orbitals Type each non-zero element in the format '*met*1,*met*2:*value*', separate adjacent matrix elements by ';' or '|'.
Example: `V_cf = (3,3:1.8 | 4,4:1.8)`.

`R_MM` :

Coulomb interaction integrals on metal orbitals in the ground state (in the form '*index,value*'; separate adjacent offsets by ';' or '|'.)
Example: `R_MM = (0:4.0 | 2:8.10 | 4:5.08)`.

`change_factor_RMM` :

For an excited state of the metal electron the Coulomb iteraction itegrals can be different from those in the ground state because the overall electronic structure is different in the excited state. This change is described by parameter `change_factor_RMM`: `R_MM`(excited)=`R_MM`(ground state) · `change_factor_RMM`. Default is 1.0 for all metal orbitals. Example: `change_factor_RMM = (0:1.05 | 2:1.04 | 4:1.08)`.

`R_CM` :

Coulomb interaction integrals of metal orbitals with the core configuration. Type each integral in the form '*index,value*'; separate adjacent offsets by ';' or '|'.
Example: `R_CM = (0:-4.0 | 2:-8.0)`.

`Ex_CM` :

Exchange integrals of metal orbitals with the core configuration.
Example: `Ex_CM = (0:0.0 | 2:0.27)`.

Program mode; which spectra are to be calculated

`calc_only_ground_state` :

Calculate only ground state, no spectra. Allowed values are 0 (= off) and 1 (= on). Default value is 0.

`calc_leftright_polarized_spec` :

Calculate X-ray absorption spectra with left and right circular X-rays as well as the difference spectrum of the two (which is measured in X-ray dichroism experiments). Allowed values are 0 (= off) and 1 (= on). Default value is 0.

`calc_xy_circ_spec` :

Calculate X-ray absorption spectra with linear polarized X-rays in x- and y-direction. Allowed values are 0 (= off) and 1 (= on). Default value is 0.

A.3 Program package "sse"

A.3.1 Program files and documentation

Program package `sse` consists of 4 groups of files:

1. main `sse` program files:

`sse_main.C`	main file for project '`sse`'
`sse_print.C`	main file for output creating program '`sseprint`'
`sse.h`	main classes `SSERun`, `SSESim`, `SSEMeasurements`
`sse_.cc`	implementations and template instantiations for '`sse.h`'
`sse_aux.ih`	template implementations for some auxiliary classes
`sse_init.ih`	functions for initializing main `sse` classes and file I/O
`sse_upd.ih`	functions performing QMC update steps (free world-line update, diagonal update, loop update, Green function measurements)
`sse_meas.ih`	functions performing measurements of observables
`sse_sim.ih`	functions for scheduling and load balancing on parallel computer architectures; formatted output of results
`sse_chk.ih`	functions for checking, debugging and protocol writing
`sse_usermeas.h`	allows users to define new observables to be measured

2. Documentation, parameters and settings:

readme.sse	list of all program files of package sse
sse_parm.txt	description of allowed parameters for parameter file
debug_protocol.h	debug tools: macros and functions for protocol writing, range checking and memory allocation checks.
sse_options.h	adjust sse to your needs via compiler flags
Makefile	makefile for creating programs 'sse' and 'sseprint' by typing 'make sse' and 'make sseprint'
make.inc	auxiliary file for 'Makefile': compiler, linker and library paths. This file probably has to be modified when transfering sse to a new computer system

3. Files of auxiliary module 'Lattice' implementing finite lattices:

my_sort.h	some sorting routines needed in lattice classes
lattice.h	main lattice concepts 'Lattice', 'Lattice_Fourier' and 'DisorderedLattice'
lattice.ih	implementation of template functions for 'lattice.h'
lattice_.cc	non-template implementations and template instantiations for 'lattice.h'
lattice_1D.h	some 1-dimensional realizations of the general 'Lattice' concept: dimer, chain and ladder lattices
lattice_1D.ih	implementation file for 'lattice_1D.h'
lattice_2D.h	some 2-dimensional realizations of the general 'Lattice' concept: rectangle, square, tilted square and bilayer lattices
lattice_2D.ih	implementation file for 'lattice_2D.h'
lattice_3D.h	some 3-dimensional realizations of the general 'Lattice' concept: cubic and multilayer latices
lattice_3D.ih	implementation file for 'lattice_3D.h'

4. Files of auxiliary modules 'SiteOccup' and 'Model', implementing orbitals, particles and many-particle Hamiltonians:

models.h	main concepts 'SiteOccup', 'SiteOccup_Manager, 'Model'
models.ih	implementation of template functions for 'models.h'
models_.cc	non-template implementations and template instantiations for 'models.h'
site_oc.h	some realizations of concepts 'SiteOccup' and 'SiteOccup_Manager'
site_oc.ih	implementation file for 'site_oc.h'
hubbmodel.h	Hubbard-like realizations of concept 'Model'
hubbmodel.ih	implementation file for 'hubbmodel.h'
spinmodel.h	Heisenberg-like realizations of concept 'Model'
spinmodel.ih	implementation file for 'spinmodel.h'

Additionally, sse needs the libraries 'Alea' and 'Osiris' [201] which are not part of the sse package itself.

A.3.2 Auxiliary concepts 'Lattice', 'SiteOccup' and 'Model'

One of the main strategies when designing `sse` was to separate the simulation technique from the physical specification of the studied system. To achive this goal the program was split into 5 independently usable object-oriented modules (see Fig. A.1):

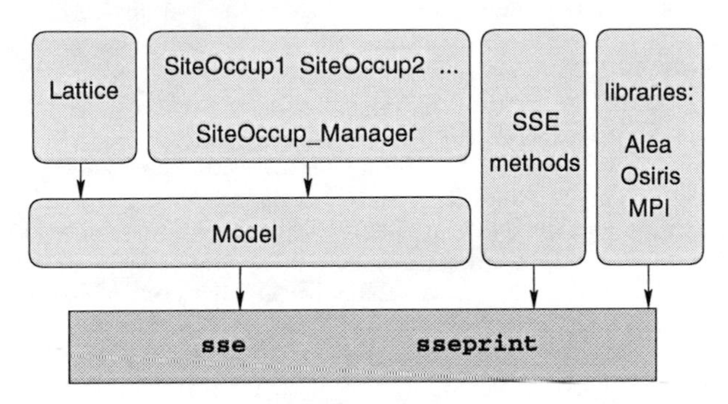

Figure A.1: Basic functional modules of a lattice simulation problem in many-particle physics

- **module 'Lattice'** contains all information on the system's geometry, on boundary conditions, symmetries, crystal momenta and interacting site pairs.
- **module 'SiteOccup'** provides concepts defining orbitals and their possible occupations by certain particles. Other concepts serve to administrate all different orbital and particle types of the physical system to be simulated.
- **module 'Model'** contains concepts defining Hamiltonians, i.e. all energetic interactions between the particles of a given system.
- **module 'SSE'** contains the SSE simulation algorithm itself with its main classes '`SSERun`' (for running one QMC process consisting of alternating update and measuring steps), '`SSEMeasurements`' (for storing the measured data and performing statistics and binning analysis) and '`SSESim`' (responsible for coordinating the various `sse` runs, for load balancing on parallel systems and for generating output files).
- **Module 'Parallelization'** consists of the libraries 'Alea' and 'Osiris' [201], which internally use library routines provided by the message passing interface (MPI) [202] software tool kit.

A.3.3 Compiler options and program modes

Program mode settings which have to be known at compile time are communicated to `sse` via the header file '`sse_options.h`'. This file contains 2 groups of settings. In the first group a lattice geometry, one or more particle and orbital types and the system's Hamiltonian have to be defined. The second group contains additional compiler switches serving to give sse additional hints on special properties of the simulated system; these hints help sse to generate optimally performing code for each physical system.

Defining lattice geometry and Hamiltonian

As an example of what the first group of settings might look like we suppose that we want to simulate a spin-1/2 Heisenberg model with some non-magnetic impurities on a square lattice with periodic boundary conditions. The Heisenberg site-site interactions $\boldsymbol{J S_i S_j}$ are to act on nearest neighbors $\langle i,j \rangle$ and – with a different interaction constant $J' \neq J$ – on next-nearest neighbors $\langle\langle i,j \rangle\rangle$. Our aim is to measure the system's energy, its susceptibility and the $\boldsymbol{k}$-dependent dynamical spin correlation function $\langle S_z(\boldsymbol{k},\omega)\, S_z(\boldsymbol{k},0)\rangle$. The class descriptions in file 'lattice_2D.h' tell us that Square_Lattice_PBC<IA_RANGE> and Square_Lattice_PBC_Fourier<IA_RANGE> with template parameter IA_RANGE = 2 (because the nearest and next-nearest neighbor bonds are active) are suitable lattices for this purpose. We choose the second variant because our class needs some knowledge about Fourier transformations to calculate $\langle S_z(\boldsymbol{k},\omega)\, S_z(\boldsymbol{k},0)\rangle$ from the $\langle S_z(\boldsymbol{r},\tau)\, S_z(\boldsymbol{r},0)\rangle$ measurements performed by sse. Going through file 'site_c.h' we find out that the site occupation class Spin_SiteOccup<TWO_S> with template parameter TWO_S = 1 is suitable to model the normal spin-1/2 sites of our system, whereas the non-magnetic impurities can be described by class Empty_SiteOccup. Furthermore, we have found out that the file 'site_c.h' also provides a matching site occupation manager: the class Impurity_SiteOccup_Manager<SiteOccup1,SiteOccup2>. We still need a suitable Hamiltonian class; header file 'spinmodel.h' seems promising for this purpose. Indeed, we find class ExtendedSpinModelWithImpurities<SiteOccup_Mgr,Lattice,Mode> and note that it matches our needs if we insert the previously found site occupation manager and lattice class as the first two template arguments and if we choose Mode = 0 (since we do not have hopping processes in our model). Now we can communicate our selections to sse by writing the folowing lines into 'sse_options.h':

```
#define _LATTICE_ Square_Lattice_PBC_Fourier<2>
#define _SITEOC_1 Spin_SiteOccup<1>
#define _SITEOC_2 Empty_SiteOccup
#define _MANAGER_ Impurity_SiteOccup_Manager<_SITEOC_1,_SITEOC_2>
#define _MODEL_   ExtendedSpinModelWithImpurities<_MANAGER_,_LATTICE_,0>
```

Compiler flags for optimizing performance

In the second part of file 'sse_options.h' one can activate or deactivate the following flags in order to optimally adapt sse to the currently studied system's properties:

#define HAS_MINUS_SIGN_PROBLEM :

Must be activated if simulating a Hamiltonian with minus sign problem (i.e. if the Hamiltonian contains nondiagonal interaction vertices with positive energy). For other Hamiltonians it should be deactivated since it produces slower code. Before activating this flag one should check whether the sign problem can be removed by local gauge transformations (see section 3.1.1 for details).

#define MAX_ALLOWED_PATHS_THROUGH_VERTEX :

This flag can take the values 1, 4 or 8; it defines an upper boundary for the number of allowed vertex modifications for a given interaction vertex and incoming leg. 4 is sufficient for most Hamiltonians and allows for slightly faster loop updates than 8. 1 can only be set if the bounce path has been optimized away (see section 3.1.1) and if there is only 1 possible non-bounce path through each vertex type. To test whether this is true one can tell `sse` to print a path information protocol by setting `PRINT_PATH_INFO_AND_EXIT=1;` in the parameter file.

`#define LOOP_ALWAYS_CLOSES_AT_STARTLEG` :

Activating this flag allows for faster loop update code if open loops are impossible for the chosen physical model. This is normally the case if there are no allowed vertices changing the particle number by ± 1.

`#define SWITCH_OFF_ANTITRAPPING` :

Per default `sse` aborts loop updates if their lengths exceed a certain threshold (which can be adjusted in the parameter file). This is done to avoid trapping in "monster" loops, in which the loop head never finds its way back to the initially inserted disturbation of the world-line configuration. If one wants to switch off that feature (which might increase speed by some percents in situations in which trapping prooves to be not a problem) one has to activate this flag.

`#define CANONICAL_MODE` :

This flag decides whether the particle number of certain particle types can be fixed during an `sse` simulation. Allowed values are 0, 1 and 2, 0 being the default mode in which all particle numbers float freely. 1 allows to keep the particle number of one particle type constant, 2 permits to fix the numbers of more than one particle type. 0 produces the fastest code, 2 the slowest.

`#define TDGF_RAW_MEA_SIZE` :

Defines a buffer size for raw Green functions measurements (which are recorded during the loop update). The optimum size has to be determined for each computer system separately. Too large buffers produce many cache misses, too small buffers stop the loop update process very often for transforming the recorded data into imaginary time and cleaning the buffer. The best value should lie somewhere in between 1000 and 500000. (Note that each raw mesurement element needs `(2*int+double+pointer)` of memory.)

`#define SSE_COUNTER` :

This flag can be activated for benchmarking or profiling purposes. It makes `sse` count the number of elementary update steps of a simulation. The result is then written into the output file created by `sseprint`. Elementary update steps are: a) visiting one diagonal or empty vertex during the diagonal update; b) performing a free world-line update on one site; c) entering the loop update and choosing a start change; d) performing one loop step from vertex to vertex in the loop update.

`#define LINK_DIAGS_IMMEDIATELY` :

Link newly created non-empty vertices into the vertex web immediately after their creation. This is slightly faster than the standard mode (which first performs all creations, then all linkings) for high temperatures.

A.3.4 Program parameters

Overview

`sse` needs a parameter file containing information on the system's geometry, the involved particle types and their energetic interactions as well as the observables to be measured. The syntax for starting `sse` is "`sse <` *param_file*" or "`sse -p` *param_file*". Optionally, one can add "`-T` *time*; here, *time* is the time interval (in seconds) after which `sse` checks whether the termination condition is fulfilled and the simulation can be stopped.
The parameter file should be a simple unformatted text file containing a series of rules of the form "*param_name* `=` *value*`;`". One parameter file can be used to start several simulations with different parameter settings; therefore the file should start with a series of rules to be valid for all simulations, followed by one specific rule set for each simulation, each of them enclosed into a pair of braces {...} in order to mark that the enclosed rules apply only to one specific simulation.
The parameters recognized by `sse` can be grouped into 8 subsets:

- Protocol writing and checking flags
- Fine-tuning and optimizations of the update and measurement process
- Initialization and termination conditions
- Automatic result file naming
- Thermodynamics: temperature and ensemble (grand-canonical or canonical)
- Selection of observables to be measured
- Properties of the crystal lattice
- Properties of particles and Hamiltonian

Each of these subsets will be described in more detail in the following subsections.

Protocol writing and performing checks

`PRINT_LATTICE_AND_MODEL_PROTOCOL_AND_EXIT` :

When simulating a new Hamiltonian for the first time one might want to check whether lattice geometry and particle properties are modeled correctly. Activating this flag makes `sse` call the self-protocol routines of the classes '`Lattice`', '`SiteOccup_Manager`' and '`Model`'. Then `sse` stops.
Allowed values: 0 (= off) and 1 (= on). Default value is 0.

`PRINT_SSE_VERTEX_TYPES_AND_EXIT` :

Having checked geometry and particles one might ask whether the elementary interactions are modeled correctly. Activating the this flag makes `sse` write a list of all interaction vertices (before and after shifting diagonals' energies) to the protocol file defined in `sse_options.h` Then `sse` stops.
Allowed values: 0 (= off) and 1 (= on). Default value is 0.

`PRINT_PATH_INFO_AND_EXIT` :

After having checked the interaction vertices one might be interested in how many path choices there are for all vertex types v and incoming legs l. If this number is smaller than 2 or at least smaller than 5 for all (v, l) combinations, the compiler flag `MAX_ALLOWED_PATHS_THROUGH_VERTEX` in file `sse_options.h` can be used to build optimized code. If above flag is activated sse prints a path info list and exits.
Allowed values: 0 (= off) and 1 (= on). Default value is 0.

`WRITE_MEMORY_OCCUPATION_PROTOCOL` :

This option makes `sse` create a detailed list showing how much memory the main objects '`SSERun`' and '`SSESim`', their parent objects and their member arrays are needing.
Allowed values: 0 (= off) and 1 (= on). Default value is 0.

`WRITE_PROFILING_INFO` :

Create a list showing how much computation time of a simulation has been spent in different parts of `sse` (diagonal and loop update, various measurements, Fourier transformations, Green function transformations).
Allowed values: 0 (= off) and 1 (= on). Default value is 0.

`MODEL`,
`LATTICE` :

These flags serve to make sure that the currently used `sse` code has been compiled for the same model and lattice for which the current parameter file is intended. If not, one gets an error message and the program aborts.
Allowed values: all strings `"..."`. Default is `""`(= no check performed).

Fine-tuning and optimizations of the update and measurement process

`LOOPSTEPSFACTOR` :

increases/decreases the number of loop update steps per sweep (i.e. between two adjacent measurements): the number of loop updates is chosen so that on average '`LOOPSTEPSFACTOR*cutoff_L`' vertices per sweep are traversed.
Allowed values: any positive real number. Default is 1.

`CUTOFF_FACTOR` :

The cut-off number in `sse`'s series expansion of the partition function is determined automatically during the thermalization process of `sse`. In some 'pathological' cases this automatic choice might proove not big enough afterwards and `sse` somewhen runs out of empty vertices. In this case one can manually increase the cut-off number by multiplying the automatically determined value with the factor '`CUTOFF_FACTOR`'.
Allowed values: any positive real number. Default is 1.

`MAX_LOOPLENGTH` :

Flag defining the loop length threshold above which a loop is aborted and the original state of the world lines is restored. If one wants to switch off loop aborting the flag '`#define SWITCH_OFF_ANTITRAPPING`' has to be activated in file `sse_options.h`.
Allowed values: integers larger than 1. Default is $50 \cdot \beta \cdot N_{sites}$.

`START_INTERPOLATING` :

In order to perform time dependent measurements as fast as possible the class `SSERun` internally stores tables of "integrated weight distributions" $\sum_{m=0}^{L} \binom{L}{m} t^m (1-t)^{L-m}$. (with $t = \tau/\beta$). It is sufficient to store only data from the range $[m1, m2]$ in which the single weight coefficients are larger than $10^{-7}...10^{-8}$, which corresponds to an interval of about $10 \cdot \sigma$ (with σ being the standard deviation of the weight distribution). As these data have to be stored for all imaginary time points one can run out of memory if the cut-off number L is very large. In that case the "integrated weight function" can be approximated by 6 adjacent polynomials (as described in section 3.1.5) in order to reduce memory consumption. Above flag is the σ value above which this approximation by polynomial interpolation starts.
Allowed values: real numbers larger than 6. Default is 40.

`OFFSET_E` :

perform a larger energy shift of diagonal interaction elements than the one chosen by `sse` automatically. This sometimes gives better autocorrelation times at the cost of increasing $\langle n_{vertices} \rangle$ and the cut-off number L.
Allowed values: positive real numbers. Default value is 0.

`ALWAYS_KEEP_BOUNCE_PATH` :

For some physical systems the "bounce" paths can be optimized away. `sse` automatically detects those systems and omits the bounce paths. Sometimes, however, the omission of the bounce paths can produce vertices with no valid path at all. In this case `sse` aborts with a message asking to switch on this flag.
Allowed values: 0 (= off) and 1 (= on). Default value is 0.

`PERFORM_FREE_WORLD_LINE_UPDATE` :

Enable or disable `sse`'s free world-line update (in which the occupation of those world-lines that are not connected to any interaction vertex at all is flipped randomly).
Allowed values: 0 (= enabled) and 1 (= disabled). Default is 1.

VERTICES :

For testing one might want to fix the number of non-empty vertices to a constant; should not be activated in production simulations!
Allowed values: 0,1,2,... Default is 0 (= no restriction).

REF_SITES_IN_EQUAL_TIME_CORR_MEASUREMENTS
REF_SITES_IN_TIME_DEP_CORR_MEASUREMENTS
REF_SITES_IN_SUM_TIME_DEP_CORR_MEASUREMENTS :

The density-density and S_z-S_z correlation measurements within sse repeatedly choose a randomly selected reference site and measure all correlations $C(r,\tau)$ with respect to this site. Above flags determine how many of those cycles of defining a reference site and measuring are performed per sweep.
Allowed values: integers larger than 0. Default values are 5.

Initialization and termination conditions

START_WORLD_LINE_CONFIG_FILE :

Per default each sse run starts with an empty vertex string and with a randomly filled start state. Alternatively, sse can be told to start all runs with a certain configuration (consisting of start state and vertices) which has previously been stored in a dump file.
Allowed values: strings "...". Default is "" (= start from random state).

END_WORLD_LINE_CONFIG_FILE :

Create a dump file at the end of the simulation which contains the last configuration of the "master run" (on node==0).
Allowed values: strings "...". Default is "" (= do nothing).

SEED :

Define a seed value for the sse's internal random number generator.
Allowed values: non-negative integers. Default value is 0.

THERMALIZATION :

Define the number of thermalization sweeps before starting measurements. In order to start measurements from a state in thermodynamic equilibrium one should make sure that THERMALIZATION is larger than 30..50 times the autocorrelation times resulting from the binning analysis of all measured quantities.
Allowed values: positive integer values. Default is 500.

SWEEPS :

Define the total number of sweeps.
Allowed values: positive integer values. Default is 0.

STOP_AT_ENERGY_PER_SITE_ERROR :

Sometimes one does not want to stop after a fixed number of sweeps but rather when the measured values' statistical errors are sufficiently small. In this case above flag defines an error threshold for the observable "energy per site"; the simulation then stops when the measured error becomes smaller than this threshold *or* when SWEEPS is reached.
Allowed values: non-negative real numbers. Default is 0 (= criterion disabled).

Automatic result file naming

`FILE_NAMING_PARAMETERS`
`FILE_NAMING_TEXTS` :

These two parameters are useful for automatic output file naming when using program '`sseprint`'. If, for example, the parameter file contains the settings "`U=5;VOLUME=100;T=0.5;t=2;mu0=1.;`", then with the two lines above `sseprint` produces the output file '`hcb_vol100_T0.5_mu1`'. If a file with this name already exists in the current directory, further files with the same parameter settings are named '`hcb_vol100_T0.5_mu1_2`', '`hcb_vol100_T0.5_mu1_3`' and so on. This guarantees that no files are overwritten unwillingly.
Allowed values: strings `"..."`. Default: `""` (standard output file name `sse_result`).

Thermodynamics: grand-canonical or canonical ensemble

`T` :

Define the temperature for the simulation.
Allowed values: positive real numbers. Default is 1.

`PARTICLE_NUMBERS` :

Per default `sse` is a grand canonical method. However, one can selectively switch to canonical measurements via this flag (Note: compiler switch `CANONICAL_MODE` must be set to a value larger than 0 in '`sse_options.h`').
Example: '`PARTICLE_NUMBERS="-1;5;-1;5"`' means: particle numbers of particle types 0 and 2 can float freely, particle numbers of types 1 and 3 are fixed to the value of 5.
Allowed values: strings `"..."` containing a non-negative integer or -1 for each particle type, separated by semicolons. Default is `""` (= all particle numbers can float freely).

Activate the quantities to be measured

`FIRST_TAU_DIV_BETA`
`LAST_TAU_DIV_BETA`
`DELTA_TAU_DIV_BETA` :

Indicate imaginary time points in units of β (start, stop, step). last value must not exceed $\beta/2$.
Allowed values: real numbers between 0 and 0.5. Defaults are 0 (= no imaginary times specified).

`R_POINTS` :

Define some selected r-points for which r-dependent correlation and Green function measurements are to be performed. There are two alternatives:
`R_POINTS="all"` activates all topologically different allowed distances
`R_POINTS="(x1,y1,z1); (x2,y2,z2); ..."` activates only the individually given r-points.

`K_POINTS` :

Define some selected k-points for which Fourier transforms from r-space are to be done for all activated k-dependent measurements. There are three allowed alternatives:
`K_POINTS="all"` activates all topologically different allowed k-points.
`K_POINTS="standard path"` activates k-points on the standard path through the Brillouin zone defined in '`Lattice`' class.
`K_POINTS="(kx1,ky1); (kx2,ky2); ..."` activates individually given k-points. The k-points can be given as multiples of π (e.g. `"(3pi/5,pi)"`) or as integers equal to $k \cdot n_{momenta}/(2\pi)$.

`MEASURE_PARTICLE_NUMBERS`
`MEASURE_OCCUPATIONS`
`MEASURE_WINDINGS`
`MEASURE_MAGNETIZATION`
`MEASURE_HEAT_CAPACITY`
`MEASURE_STRUCTURE_FACTORS`
`MEASURE_CANONICAL_ENERGIES`
`MEASURE_N_N_SUSCEPTIBILITY`
`MEASURE_SZ_SZ_SUSCEPTIBILITY`
`MEASURE_AF_SZ_SZ_SUSCEPTIBILITY` :

Select quantities to be mesured and stored by `sse`. The names should be self-explaining.
Allowed values: 0 (= off) and 1 (= on). Defaults are 0.

`MEASURE_SUM_SZ_SZ_CORRELATIONS`
`MEASURE_SUM_AF_SZ_SZ_CORRELATIONS`
`MEASURE_SZ_SZ_CORRELATIONS`
`MEASURE_SUM_TIME_DEP_SZ_SZ_CORRELATIONS`
`MEASURE_SUM_TIME_DEP_AF_SZ_SZ_CORRELATIONS`
`MEASURE_TIME_DEP_SZ_SZ_CORRELATIONS` :

Activate r-dependent S_z-S_z correlation measurements.
Allowed values: 0 (= off) and 1 (= on). Defaults are 0.

MEASURE_SUM_N_N_CORRELATIONS
MEASURE_N_N_CORRELATIONS
MEASURE_SUM_TIME_DEP_N_N_CORRELATIONS
MEASURE_TIME_DEP_N_N_CORRELATIONS :

Measure r-dependent density-density correlations for some selected particle types. Allowed values: 0 (= off) and 1 (= on). Defaults are 0.

PARTICLE_TYPE_IN_N_N_CORRELATIONS :

Activate certain particle types for which r-dependent density-density correlation measurements are to be performed.
Allowed values: strings "..." one or more particle type ID's separated by ',' or ';'. Default is "" (= no active particle types).

PARTICLE_TYPE_IN_HISTOGRAMS
PARTICLE_NR_HISTOGRAM_BINS
PARTICLE_NR_HISTOGRAM_START_VALUE
PARTICLE_NR_HISTOGRAM_STEP
WINDING_HISTOGRAM_BINS
WINDING_HISTOGRAM_START_VALUE
WINDING_HISTOGRAM_END_VALUE
PARTICLE_TYPE_IN_HISTOGRAMS :

Create histograms for particle numbers or windings: PARTICLE_TYPE_IN_HISTOGRAMS is a string containing the particle type-IDs of the particles for which histograms are to be created (default is "0").
..._HISTOGRAM_STEP is the width of each bin.
..._HISTOGRAM_START_VALUE is the lower boundary of the first bin. The last bin ends at ..._START_VALUE+...STEP*..._BINS.
Default value for the particle types is "0", for the start values 0, for the step widths 1 and for the numbers of bins 100.

MEASURE_GREENS_FUNCTIONS
MEASURE_CHANGE0
MEASURE_CHANGE1
MEASURE_CHANGE2
MEASURE_CHANGE3... :

Perform equal time Green functions measurements. If "=1", perform measurements for all allowed change pairs that can be formed with changes 'MEASURE_CHANGE0', 'MEASURE_CHANGE1' and so on as first (i.e. right) change.
Allowed values: 0 or 1 for MEASURE_GREENS_FUNCTIONS (default is 0), strings "..." containing one valid change name for the other parameters (defaults are "").

MEASURE_TIME_DEP_GREENS_FUNCTIONS
MEASURE_TIME_DEP_CHANGE0
MEASURE_TIME_DEP_CHANGE1... :

The same for time-dependent Green functions measurements.

MEASURE_COMBI_MEA :

This item allows to construct a quantity to be measured which is composed of a sum of several change pairs. Example: "GF(k,tau): <a2,a2> +<a2+,a2> +<a2,a2+> +<a2+,a2+>" means that the sum of 4 time-dependent Green functions in k-space is to be measured, each of the 4 terms being composed of creators/annihilators of particle type 2. Other possibilities are "GF(r,0): ...", "GF(r,tau): ...", "GF(k,0): ...". (Note: in the example above, MEASURE_TIME_DEP_CHANGE0="a2" and MEASURE_TIME_DEP_CHANGE1="a2+" must be activated, otherwise one gets an error message because the program can combine only activated change pairs.)

STORE_R_DEP_N_N_CORRELATIONS
STORE_R_DEP_SZ_SZ_CORRELATIONS
STORE_R_DEP_TIME_DEP_N_N_CORRELATIONS
STORE_R_DEP_TIME_DEP_SZ_SZ_CORRELATIONS
STORE_R_DEP_GREENS_FUNCTIONS
STORE_R_DEP_TIME_DEP_GREENS_FUNCTIONS
STORE_K_DEP_N_N_CORRELATIONS
STORE_K_DEP_SZ_SZ_CORRELATIONS
STORE_K_DEP_TIME_DEP_N_N_CORRELATIONS
STORE_K_DEP_TIME_DEP_SZ_SZ_CORRELATIONS
STORE_K_DEP_GREENS_FUNCTIONS
STORE_K_DEP_TIME_DEP_GREENS_FUNCTIONS :

These types of parameters can be given for all r-dependent measurements; they indicate whether the r-dependent measurements and/or their Fourier transforms into k-space are to be stored by sse.
Allowed values: 0 (= off) and 1 (= on). Defaults are 0.

Define properties of the crystal lattice

VOLUME :

Define lattice volume.
Allowed values and defaults depend on chosen Lattice class.

REINITIALIZE_INTERVAL :

This flag is intended for lattices/models with randomly distributed impurities or random interactions. It indicates that after a certain number of sweeps the lattice/model is to be (randomly) re-initialized, followed by a new thermalization phase before measurements are continued.
Allowed values: non-negative integers. Default is 0 (= no re-initializations are performed).

... (other parameters, depending on chosen Lattice class)

Define properties of particles and Hamiltonian

Interaction constants, depending on chosen model, e.g.: t=1; U=4; mu=0.6;

A.3.5 User-defined enhancements of sse

One of the guiding principles in designing `sse` was to create a modular component structure which can be used to solve wide classes of lattice simulation problems. Therefore it is unlikely that any possible application of `sse` can be foreseen and taken into account by the original implementation. A better approach is to provide future users of `sse` the possibility to easily enhance `sse`'s capabilities and to adapt the code to their specific needs. Two possible lines of enhancements are particularly supported by the present implementation:

1. The header files '`lattice.h`', '`site_oc.h`' and '`models.h`' provide documented and commented class interfaces for `Lattice`, `Site_Occupation` and `Model` classes. Together with the existing specialized classes (which can serve as examples) this documentation should allow the user to quickly write his own specialized `Lattice`, `Site_Occupation` and `Model` classes if he wants to simulate a lattice geometry or a physical model which is not covered by the standard classes coming with `sse`.

2. The user can add self-defined observables to be measured and recorded by `sse`. To this purpose the file '`sse_usermeas.h`' contains a class '`UserDefined_Measurements`; and some commenting text. In this class the user can easily incorporate his own observables. Per default, an empty version of `UserDefined_Measurements` is activated. Additionally, '`sse_usermeas.h`' contains an example version of the class containing one additional observable. This second version is commented out per default.

A.4 Program package "QM"

The program package `QM` facilitates performing algebraic manipulations of quantum mechanical expressions using the formalism of Second Quantization. It defines objects like "operators", "kets", "bras", "scalar products" or "commutators" and a set of functions combining them like non-commutative concatenation, inversion, Hermitean adjunction and so on. A large set of rules, which can be flexibly activated or deactivated, allows both to perform fully automatic normalization and simplification of large terms as well as user-controlled step-by-step transformations. The program provides occupation number base states for eigenvalue problems and a large set of meta programs and auxiliary routines permitting to easily enlarge the program's capabilities and to introduce new user-defined operators and other quantum mechanical objects.

A.4.1 Program files and documentation

A.4.2 Purpose and basic ideas

Program `QM.nb` has two main purposes, namely

- fully automatic term transformation and simplification (e.g. to proove that a certain commutator is zero or to solve an eigenvalue problem analytically).

- user-controlled step-by-step transformation of terms from an initial form into a form preferred by the user.

These two aims are somehow contradictory: the first one requires a large set of definitions and replacement rules to be applied automatically whenever possible, while those automatic replacements are not at all wanted in the second case, where at each step only one single replacement rule is to be applied exactly once. Another problem to be kept in mind is that modeling the whole formalism of Second Quantization requires such a large set of definitions and rules that Mathematica is considerably slowed down if all these rules are activated simultaneously. To account for these two aspects `QM.nb` consists of many relatively small packages of rules that can be activated and deactivated independently of each other. Each package can be activated in several ways:

- `On[`*command*`]` activates the definitions associated with the command *command*, i.e. the replacements are automatically performed whenever possible.
- `Off[`*command*`]` deactivates the definitions associated with the command *command*, i.e. the replacements are not performed any more.
- *command*`[...]` or `...//`*command* activates the corresponding definitions as temporary replacement rules (applied only one time, not always).

User defined rule sets can be added using the routines

- `createRuleSet[`*name*`,`*txt*`,`*incompat*`,`*needed*`,`*unprotec*`,`*rules*`,`*nParams*`:0,`*delayed*`:1]` (define a new package of rules and call it *name*)
- `createCompiledRuleSet[`*name*`,`*txt*`,`*incompat*`,`*needed*`,`*rules*`]` (named package of rules in precompiled form for faster evaluation)
- `createFormat[`*name*`,`*txt*`,`*incompat*`,`*needed*`,`*unprotec*`,`*rules*`]` (named package of output directives for TeX-like typographical output)

with the following argument meanings:

- *txt* is a text message to be issued if the rule set is switched on or off.
- *incompat* is a list containing the rule sets which are incompatible with the present one and which are automatically switched off if this one is activated, whereas
- *needed* is just the contrary: a list of rules to be activated automatically together with the current package.
- *unprotec* is a (possibly empty) list of standard Mathematica commands like `Plus` or `Times` for which new rules are defined in the current set and whose protection tag therefore has to be removed temporarily.
- *rules* is a list of strings "*lhs*`:=`*rhs*" containing the replacement rules forming the present set.
- *nParams* (default: 0) is the number of parameters for parameterized rule sets.
- *delayed* (default: True) controls whether definitions and rules are implemented as 'delayed' (`:=` and `:>`) or 'immediate' (`=` and `->`).

A description of all packages coming with program `QM.nb` will be given in section A.4.6.

A.4.3 Notations

QM.nb uses the following notation conventions for building up composed expressions of quantum mechanical operators, states, prefactors, numbers and so on:

Input form	Output form	Description
ket[x]	$\vert x\rangle$	q.m. state ("ket")
bra[x]	$\langle x\vert$	adjoint of q.m. state ("bra")
ket[0], bra[0]	$\vert 0\rangle$, $\langle 0\vert$	vacuum state
bracket[x,y]	$\langle x, y\rangle$	scalar product ("bra" times "ket")
commutator[x,y]	$[x, y]_-$	commutator
antiCommutator[x,y]	$[x, y]_+$	anticommutator
Op1 ** Op2	*Op1 Op2*	non-commutiative multiplication (of operators, kets, bras)
aj[Op]	$Op^\dagger$	Hermitean adjunction
inv[Op]	Op^{-1}	inversion
delta[i,j]	$\delta_{i,j}$	Kronecker-δ
hBar	$\hbar$	Planck's constant $\hbar$

A.4.4 Predefined operators

The following operators are predefined (and their properties and commutator relations are known to the system):

Fermion constructors and destructors

Input form	Output form	Description
c[i,s]	$c_{i,s}$	destructor of a spin-s Fermion at site i
cDag[i,s]	$c_{i,s}^\dagger$	constructor of a spin-s Fermion at site i
n[i,s]	$n_{i,s}$	number of spin-s Fermions at site i
n[i]	n_i	$n_i = n_{i,\uparrow} + n_{i,\downarrow}$
cHat[i,s]	$\hat{c}_{i,s}$	modified destructor $\hat{c}_{i,s} = c_{i,s}(1 - n_{i,-s})$
cHatDag[i,s]	$\hat{c}_{i,s}^\dagger$	modified constructor $\hat{c}_{i,s}^\dagger = c_{i,s}^\dagger(1 - n_{i,-s})$
dHat[i,s]	$\hat{d}_{i,s}$	modified destructor $\hat{d}_{i,s} = c_{i,s}n_{i,-s}$
dHatDag[i,s]	$\hat{d}_{i,s}^\dagger$	modified constructor $\hat{d}_{i,s}^\dagger = c_{i,s}^\dagger n_{i,-s}$

Spin operators

Input form	Output form	Description
Sx[i]	S_i^x	spin in x-direction at site i
Sy[i]	S_i^y	spin in y-direction at site i
Sz[i]	S_i^z	spin in z-direction at site i
Splus[i]	S_i^+	spin raising operator: $S_i^+ = S_i^x + iS_i^y$
Sminus[i]	S_i^-	spin lowering operator: $S_i^- = S_i^x - iS_i^y$

Triplet boson operators

Input form	Output form	Description
`tx[i]`	$t_{i,x}$	x-triplet at site i
`ty[i]`	$t_{i,y}$	y-triplet at site i
`tz[i,-1/0/1]`	$t_{i,\downarrow}, t_{i,z}, t_{i,\uparrow}$	3 components of z-triplet at site i
`tvec[i]`	$\{t_{i,x}, t_{i,y}, t_{i,z}\}$	$\{t_{i,x}, t_{i,y}, t_{i,z}\}$
`nz[i,-1/0/1]`	$n_{i,\downarrow}, n_{i,\rightarrow}, n_{i,\uparrow}$	particle number of each type of t_z-triplets at site i
`nz[i]`	$\mathcal{N}_i$	total number of all t_z-triplets at site i

SO(5) operators

Input form	Output form	Description
`t1[i]`, `t5[i]`	$t_{i,1}$, $t_{i,5}$	charge components of superspin at site i
`th[i]`, `tp[i]`	$t_{i,h}$, $t_{i,p}$	hole- and charge-pair destructors at site i
`so5[i,1...5]`	–	$t_{i,1}/t_{i,x}/t_{i,y}/t_{i,z}/t_{i,5}$
`so5vec[i]`	–	$\{t_{i,1}, t_{i,x}, t_{i,y}, t_{i,z}, t_{i,5}\}$
`nh[i]`	$\mathcal{N}_i^h$	number of hole pairs at site i
`superspin[i,a]`	–	(so5[i,a]+so5Dag[i,a])/$\sqrt{2}$
`Casimir[i,a,b]`	–	-I so5Dag[i,a] so5[i,b] + I so5Dag[i,b] so5[i,a]

A.4.5 Commands defining properties of variables

There are several commands allowing to give the system additional information on certain variables:

Command	Description
`setNonCommutative[x,y,...]`	the variables $x, y, ...$ are non-commutative with respect to `**`
`setCommutative[x,y,...]`	the variables $x, y, ...$ are commutative with respect to `**` (default for all new variables)
`setNonZero[x,y,...]`	the variables $x, y, ...$ have non-zero values
`removeNonZero[x,y,...]`	remove previous `setNonZero` command
`setUnequal[x,y]`	the variables x and y are different
`removeUnequal[x,y]`	remove previous `setUnequal` command
`assumeUnequal[x,y]`	rule set assuming that x and y are different
`setSpinIndex[x,y,...]`	the variables $x, y, ...$ are $+1/2$ or $-1/2$
`removeSpinIndex[x,y,...]`	remove previous `setSpinIndex` command
`setContainsSiteIndex[...]`	should be activated for real space q.m. operators
`removeContainsSiteIndex[...]`	remove previous `setContainsSiteIndex`

Per default, a new variable has none of the special properies given above and is taken to be `commutative` (Notice, however, that the expressions `ket[.]`, `bra[.]`, the predefined operators (`c`, `cDag`, ...) and all vectors and matrixes are non-commutative per default).

A.4.6 Replacement rule packages

Most commands in `QM.nb` provide sets of replacement and simplification rules to be used as described in section A.4.2. In general these rules are switched off per default and have to be activated when needed using the `On[.]` command. Four very fundamental rule sets, however, are activated per default: `NCBasics`, `deltaBasics`, `fermiOpBasics`, `spinOpBasics` and `qmFormat`. All commands will be described in detail in the following subsections.

General replacement rules

These rules connect the operation `**` with properties of the non-commutative operator concatenation in quantum mechanics; furthermore, they define properties of the Kronecker-δ symbol and vacuum state $|0\rangle$:

Command (Short Cut)	Description
`NCBasics` (`NCB`)	basic properties of non-commutative multiplication `**`
`NCExpand` (`NCE`)	expand sums within `**`-terms into sums *of* `**`-terms
`NCExpandAll` (`NCEA`)	expand sums within multiplicative terms (`**` and `*`) into sums *of* purely multiplicative terms
`CoefExpand` (`CoE`)	apply `Expand` to commutative prefactors of `**`-terms
`CoefSimplify` (`CoE`)	apply `Simplify` to commutative prefactors of `**`-terms
`deltaBasics` (`deB`)	basic properties Kronecker-δ
`deltaSimplify` (`deS`)	enhanced simplification rules for terms including Kronecker-δ's
`vacStateProperties` (`vSP`)	properties of vacuum state $|0\rangle$
`matrixVectorMult` (`mVM`)	rules for `**`-terms with vectors and matrices; defines cross product `NCCross[v,w]`
`qmFormat` (`qmF`)	TeX-like output format for q.m. operator terms

Replacement rules for fermion operators

Properties of the fermion constructor `cDag[i,s]` and destructor `c[i,s]` are defined here, in particular commutator relations, normal forms, and switching between `n[i,s]` and `cDag[i,s]**c[i,s]`.

Command (Short Cut)	Description
`fermiOpBasics` (`fOB`)	basic properties of $c_{i,s}$, $c^{\dagger}_{i,s}$ and $n_{i,s}$
`fermiOpSimplify` (`fOS`)	enhanced simplification rules for terms including $c_{i,s}$, $c^{\dagger}_{i,s}$ and $n_{i,s}$
`fermiOpNormalForm` (`fONF`)	conversion of Fermi operator terms (including $n_{i,s}$, $c_{i,s}$ and $c^{\dagger}_{i,s}$) into a normal form

Command (Short Cut)	Description
`separateConstrDestr` (`sCD`)	separation of Fermi constructors and destructors in `**`-terms
`nToConstrDestr` (`n2CD`)	automatic replacing of $n_{i,s}$ by $c_{i,s}^\dagger c_{i,s}$
`constrDestrToN` (`CD2n`)	automatic replacing of $c_{i,s}^\dagger c_{i,s}$ by $n_{i,s}$

Replacement rules for spin operators

properties of spin operators `Sx[i]`, `Sy[i]`, `Sz[i]`, `Splus[i]` and `Sminus[i]`; expressing spin operators in terms of fermion operators and vice versa:

Command (Short Cut)	Description
`spinOpBasics` (`sOB`)	basic properties of S_i^z, S_i^+ and S_i^-
`spinOpSimplify` (`sOS`)	enhanced simplification rules for terms including S_i^z, S_i^+ and S_i^-
`spinOpToFermiOp` (`sO2FO`)	express spin operators in terms of Fermi operators $n_{i,s}$, $c_{i,s}$ and $c_{i,s}^\dagger$
`fermiOpToSpinOp` (`fO2SO`)	find spin operator terms within other terms including Fermi operators $n_{i,s}$, $c_{i,s}$ and $c_{i,s}^\dagger$
`spinOpFermiOpSimplify` (`sOFOS`)	enhanced simplification rules for terms including both spin and Fermi operators

Replacement rules for modified fermion operators

some properties of modified fermion operators `cHat[i,s]:=c[i,s]**(1-n[i,-s])` and `dHat[i,s]:=c[i,s]**n[i,-s]`; expressing these operators in terms of fermion operators and vice versa:

Command (Short Cut)	Description
`cHatDHatBasics` (`cHDHB`)	basic properties of $\hat{c}_{i,s}$ and $\hat{d}_{i,s}$
`cHatDHatSimplify` (`cHDHS`)	enhanced simplification rules for terms including $\hat{c}_{i,s}$ and $\hat{d}_{i,s}$
`cHatDHatToFermiOp` (`cHDH2FO`)	express 'hat' operators in terms of Fermi operators $n_{i,s}$, $c_{i,s}$ and $c_{i,s}^\dagger$
`fermiOpToCHatDHat` (`fO2CHDH`)	express fermion operators $n_{i,s}$, $c_{i,s}$ and $c_{i,s}^\dagger$ in terms of 'hat' operators
`findCHatDHatOp` (`fCHDH`)	find $\hat{c}_{i,s}$ and $\hat{d}_{i,s}$ terms within other terms including Fermi operators $n_{i,s}$, $c_{i,s}$ and $c_{i,s}^\dagger$
`fermiOpCHatDHatSimplify` (`fOCHDHS`)	enhanced simplification rules for terms including both Fermi operators and $\hat{c}_{i,s}$ and $\hat{d}_{i,s}$
`spinOpCHatDHatBasics` (`sOCHDHB`)	basic properties of terms containing 'hat' and spin operators

Command (Short Cut)	Description
spinOpToCHatDHat (sO2CHDH)	express spin operators in terms of 'hat' operators
cHatDHatToSpinOp (cHDH2SO)	express 'hat' operators in terms of spin operators
cHatDHatFormat (cHDHF)	TeX-like output formatting for 'hat' operators

Replacement rules for triplet boson operators

properties of operators tx[i], ty[i] and tz[i,s]; expressing these operators in terms of fermion operators and vice versa:

Command (Short Cut)	Description
tripletOpBasics (tOB)	basic properties of triplet operators
tripletOpSimplify (tOS)	enhanced simplification rules for terms including triplet operators
tripletOpNormalForm (fONF)	conversion of **-terms containing triplet operators into a normal form
separateTzDagTz	tz[i,s]**tzDag[j,t] to tzDag[j,t]**tz[i,s]
txTyToTz	replace tx[i] and ty[i] by tz[i,1] and tz[i,-1]
tzToTxTy	replace tz[i,1] and tz[i,-1] by tx[i] and ty[i]
nzToTzDagTz	replace nz[i,s] by tzDag[i,s]**tz[i,s]
tzDagTzToNz	replace tzDag[i,s]**tz[i,s] by nz[i,s]
combineNz	combine nz[i,-1], nz[i,0] and nz[i,1] to nz[i]
tripletOpToFermiOp	express triplet operators in terms of fermion operators
nzToSz	combine two terms containing nz[i,1] and nz[i,-1, respectively, into one term containing Sz[i]
tripletFormat (tOF)	TeX-like output formatting for triplet operators

Replacement rules for SO(5) operators

properties of operators t1[i], t5[i], th[i], tp[i], so5[i,a], superSpin[i,a] and Casimir[i,a,b]; expressing these operators in terms of fermion operators:

Command (Short Cut)	Description
SO5OpBasics (SO5B)	basic properties of SO(5) operators
t1T5ToThTp	replace t1[i] and t5[i] by th[i] and tp[i]
thTpToT1T5	replace th[i] and tp[i] by t1[i] and t5[i] by th[i]
nhToThDagTh	replace nh[i] by thDag[i]**th[i]
thDagThToNh	replace thDag[i]**th[i] by nh[i]
SO5OpToFermiOp	express SO(5) operators in terms of fermion operators
SO5Format (SO5F)	TeX-like output formatting for SO(5) operators

A.4.7 Quantum mechanics on finite lattices

Finite lattices

Per default two-dimensional (2D) real-space site indexes of operators should be given in the form `site[i,j]`. This makes sure that all simplification rules for operators and occupation number states can be correctly applied. The following commands allow to use other forms for 2D site indexes:

- `easyIndexInput2D` (rule set):
  Allows 2D site indexes of fermion and spin operators to be given as lists `{i,j}` (which are automatically converted to `site[i,j]`).

- `convertSiteIndexesTo1D` (rule set):
  Replaces all `site[i,j]`-terms by `i+j*nX`.

- `convertSiteIndexesTo2D[.]` (function):
  Replaces combined site indexes `i`(as produced by `convertSiteIndexesTo1D` into standard form `site[Mod[i,nX],Quotient[i,nX]]`.

A certain fixed lattice size and/or boundary conditions can be defined via the parameterized rule sets

- `clusterSizeAndBC1D[`*nX, phiX*`]` and

- `clusterSizeAndBC1D[`*nX, nY, phiX, phiY*`]`.

Here *nX* and *nY* are the number of lattice sites in x- and y-direction and *phiX* and *phiY* describe the boundary conditions (<0: closed, $=0$: open , $]0,2\pi[$: tilted).

Occupation number base states for spin-1/2 fermions

Expressions of the form `ket[`*list*`]` or `bra[`*list*`]` concatenated with fermion or spin operators are interpreted as occupation number states if *list* has the form

$$list = \{\{n_{1,\downarrow},n_{1,\uparrow}\},\{n_{2,\downarrow},n_{2,\uparrow}\},\ldots\}.$$

If a site-dependent operator like `c[i,s]` – where `i` is either an integer or a composed object like a list or an expression with head `site` – is applied onto such a state, the function `siteNr[i]` is called to decide on which pair $\{n_\downarrow,n_\uparrow\}$ in the list the operator acts. The following rule sets are useful in this context:

- `ocStateBasics`:
  Defines the action of operators `c[i,s]`, `cDag[i,s]`, `n[i,s]`, `Sz[i]`, `Splus[i]`, `Sminus[i]`, `hopp[i,s]`, `SzSz[i]` and `SplusSminus[i]` on occupation number states.

- `ocStateBasicsFast`:
  Same functionality as `ocStateBasics`, but uses precompiled functions internally (via auxiliary rule set `ocStateCompiled`) and evaluates thus much faster.

- `ocStateSymOps1D2D`:
  Defines translations, rotations and reflections of occupation number states in 1D or 2D.

- `ocStateFormat`:
  Defines TeX-like output formats for occupation number states.

Hilbert bases with specific properties

Some specific Hilbert bases for spin-1/2 fermions can be easily created using the functions

- `createProductBase`[$N_S, n, double_oc$]:
  product base of one-site states for N_S lattice sites and n fermions; if $double_oc = 0$ double occupancies on one site are forbidden.

- `createNSzkBase`[$N_S, N_x, n, n_\uparrow, k_x, k_y, tr_x, tr_y$]:
  eigenbase of particle number, z-spin and momentum. (k_x and k_y are the wave numbers in x and y, tr_x and tr_y are the number of symmetry translations of the lattice in x and y.)

- `createNSzkPxBase`[$N_S, N_x, n, n_\uparrow, k_x, k_y, P_x$]:
  eigenbase of particle number, z-spin, momentum in x and y and reflection symmetry in x-direction. (k_x and k_y are the wave numbers in x and y, P_x is the parity under x-reflection.)

Calculating matrix elements

The matrix representation of a Hermitean operator in a given occupation number base can be computed with the two functions

- `HermOpMatrix`[*Op,baseKets,prn*:0] and

- `HermOpMatrixOcLists`[*Op,occupLists,prn*:0].

The first version takes a Hilbert base consisting of `ket[.]`'s as second argument, the second version a list of 'raw' site occupation lists without the head `ket[]` wrapped around. *prn* indicates whether or not each matrix element is to be printed on screen immediately after its caculation.

B

A fast algorithm for selective Fourier problems

Standard fast Fourier transformation algorithms (FFT) provide an efficient way to transform a set of n values $\{f(r_1), ..., f(r_n)\}$ into n values $F(k_1), ..., F(k_n)$ with $F(k_i) = \sum_{j=1}^{n} f(r_j) e^{-ir_j k_i}$. If, however, only a small subset of all $F(k_i)$ is to be calculated, these methods inevitably do "too much" and therefore show poor performance. This appendix presents an optimization technique that produces almost optimal code both for "complete" and for "selective" Fourier problems. The method works for arbitrary multi-dimensional Fourier transforms and both for real and complex Fourier coefficients.

B.1 Introduction and example problem

Imagine we simulate a two-dimensional physical system on $n \times n$ regularly distributed sites. Having measured some correlation function $f(r)$ in the n^2 accessible points r_j of real space we are now interested in the Fourier transformed function

$$F(k) := \sum_j e^{-ir_j k} = \sum_j e^{-\frac{i}{\hbar} r_j p},$$

where k denotes a wave number and $p = \hbar k$ a momentum.
From textbooks on solid state physics we know that the physically most interesting range in k-space, the first Brillouin zone [5], contains n^2 discrete accessible k-points [5]. A discrete Fourier transformation between the n^2 r-points and k-points can easily be performed using standard FFT [203] routines, e.g. the performant software package FTTW [204]. In practical problems, however, one rarely needs *all* accessible k-points. A good insight into the global behavior of $F(k)$ can be gained from regarding only the k-points on a "standard path" probing the main directions and the edge of the first Brillouin zone (see Fig. 4.1). Calculating all $F(k)$ values would contribute little new insight at the cost of massively increasing storage requirements.
Standard FFT codes, however, inevitably return $F(k)$ for *all* k-points. This inflexibility is directly connected with the divide–and–conquer scheme upon which all FFT variants are

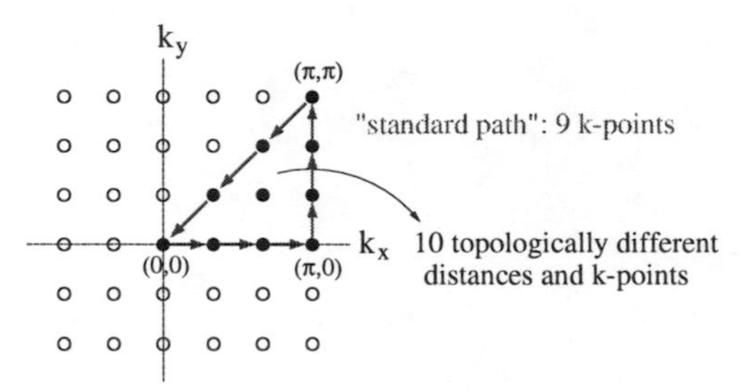

Figure B.1: A typical Fourier transform problem occuring in lattice simulations of many-body systems: a correlation function $f(r)$ has been recorded in real space on a 6×6 cluster with periodic boundary conditions (r is the distance between the two correlated lattice sites). Now one wants to know what f looks like in the Fourier-transformed space of wave numbers (or momenta) k. More precisely, one is interested in $F(k)$ for the k-points on a "standard path" exploring the first Brillouin zone. Standard FFT routines do not show optimum performance here since they neccessarily return $F(k)$ for all k-points.

based. Therefore, FFT codes do "much more than wanted" in many applications arising from solid state physics. For the example problem sketched in Fig. 4.1 the different "classical" FT and FFT approaches have the following computational complexities:

- A simple FT needs $36 \cdot 36 \cdot 2 = 2592$ complex elementary operations (additions and multiplications) to compute all $F(k)$ values and $36 \cdot 9 \cdot 2 = 648$ operations to compute only the 9 $F(k)$ values on the "standard path" of Fig. 4.1.

- FFT needs $36 \cdot \log_2(64) \cdot (2...4) \approx 360$ real or complex operations to compute all $F(k)$ points, the exact number of operations depending on implementation details. Calculating only the 9 $F(k)$ values on the "standard path" is not possible within FFT.

- A simple FT exploiting cluster symmetries and topological equivalences of distances needs $10 \cdot 9 \cdot 2 = 180$ real operations to calculate the 9 desired $F(k)$ values (since only 10 of the 36 allowed r-points are topologically different). A clever implementation of FFT might also use symmetries to reduce the computational effort; however, the gain will be much smaller than in the FT case, because exploiting symmetries makes the handling of FFT's divide–and–conquer scheme more difficult.

The basic idea to be outlined in this work is to replace FFT's "rigid" divide–and–conquer algorithm by a more flexible optimization scheme, which we will call "selective fast Fourier transform" or "SFFT" in the rest of the paper. We will show that SFFT needs only 45 operations to calculate the 9 $F(k)$ values of Fig. 4.1, i.e. only 5 arithmetic operations per k-point. For larger system sizes than 6×6 the gain with respect to the other approaches still increases.

As a starting point for introducing the ideas of SFFT we consider the Fourier coefficient matrix of our example problem in the variant that exploits symmetries. The 10 topologically different distances are

$$\{r\} = \Big\{ (0,0),\, (0,1),\, (1,1),\, (0,2),\, (1,2),\, (2,2),\, (0,3),\, (1,3),\, (2,3),\, (3,3) \Big\}$$

with multiplicities

$$\{m\} = \{\, 1,\, 4,\, 4,\, 4,\, 8,\, 4,\, 2,\, 4,\, 4,\, 1 \,\},$$

since the distance (0,1) is topologically equivalent to (0,1), (1,0), (0,-1) and (-1,0), the distance (1,1) to (1,1), (1,-1), (-1,1) and (-1,-1), and so on. Let $f_0, \ldots, f_9$ be the recorded function values for these 10 r values. The 9 k-points on the path in figure 4.1 are

$$\{\mathbf{k}\} = \Big\{(0,0), (0,\frac{\pi}{3}), (0,\frac{2\pi}{3}), (0,\pi), (\frac{\pi}{3},\pi), (\frac{2\pi}{3},\pi), (\pi,\pi), (\frac{2\pi}{3},\frac{2\pi}{3}), (\frac{\pi}{3},\frac{\pi}{3})\Big\}.$$

As the correlation function $f(r)$ has identical values for topologically equivalent r-points, the Fourier coefficients e^{-ikr} of each group of topologically equivalent r-points can be summed up into one combined Fourier coefficient

$$c_{i,j} := \sum_{r' \,\text{equiv.to}\, r_j} e^{ik_i r'} .$$

Expressed in terms of these coefficients the desired $F(k)$ values are

$$F_i = \sum_{j=0}^{9} c_{i,j}\, f_j ,$$

where F_i and f_j are short cuts for $F(k_i)$ and $f(r_j)$. The 9×10 matrix of the combined coefficients $c_{i,j}$ takes the form

k	f_0	f_1	f_2	f_3	f_4	f_5	f_6	f_7	f_8	f_9
$(0,0)$	$\mathbf{1}$	$\underline{4}$	$\mathit{4}$	$\underline{\mathbf{4}}$	8	$\mathit{4}$	2	$\underline{\mathbf{4}}$	$\underline{4}$	$\mathbf{1}$
$(\pi/3,0)$	1	3	2	1	0	-2	0	-1	-3	-1
$(2\pi/3,0)$	$\mathbf{1}$	$\underline{1}$	$\mathit{-2}$	$\underline{\mathbf{1}}$	-4	$\mathit{-2}$	2	$\underline{\mathbf{1}}$	$\underline{1}$	$\mathbf{1}$
$(\pi,0)$	1	0	-4	4	0	4	0	-4	0	-1
$(\pi,\pi/3)$	$\mathbf{1}$	$\underline{-1}$	$\mathit{-2}$	$\underline{\mathbf{1}}$	4	$\mathit{-2}$	-2	$\underline{\mathbf{1}}$	$\underline{-1}$	$\mathbf{1}$
$(\pi,2\pi/3)$	1	-3	2	1	0	-2	0	-1	3	-1
(π,π)	$\mathbf{1}$	$\underline{-4}$	$\mathit{4}$	$\underline{\mathbf{4}}$	-8	$\mathit{4}$	-2	$\underline{\mathbf{4}}$	$\underline{-4}$	$\mathbf{1}$
$(2\pi/3,2\pi/3)$	$\mathbf{1}$	$\underline{-2}$	$\mathit{1}$	$\underline{\mathbf{-2}}$	2	$\mathit{1}$	2	$\underline{\mathbf{-2}}$	$\underline{-2}$	$\mathbf{1}$
$(\pi/3,\pi/3)$	$\mathbf{1}$	$\underline{2}$	$\mathit{1}$	$\underline{\mathbf{-2}}$	-2	$\mathit{1}$	-2	$\underline{\mathbf{-2}}$	$\underline{2}$	$\mathbf{1}$

Closer inspection of this matrix shows that several columns contain almost identical coefficients – e.g. columns 0 and 9 (bold numbers), 1 and 8 (underlined numbers), 2 and 5 (italic numbers), and 3 and 7 (underlined bold numbers). The elimination of these sets of identical coefficients is the basic idea behind SFFT.

B.2 Optimizing Fourier coefficient arrays

the additions $f_0 + f_9$, $f_1 + f_8$, $f_2 + f_5$ and $f_3 + f_7$ are performed 6 times each. Hence $4 \cdot 5$ additions can be economized by introducing 4 auxiliary variables

$$f_{10} := f_0 + f_9 , \qquad f_{11} := f_1 + f_8 ,$$
$$f_{12} := f_2 + f_5 , \qquad f_{13} := f_3 + f_7 .$$

The resulting coefficient matrix including the auxiliaries then takes the form

k	f_0	f_1	f_2	f_3	f_4	f_5	f_6	f_7	f_8	f_9	f_{10}	f_{11}	f_{12}	f_{13}
$(0,0)$					8		2				1	4	4	4
$(\pi/3,0)$	**1**	3	*2*	**1**		*-2*		**-1**	-3	**-1**				
$(2\pi/3,0)$					-4		2				1	1	-2	1
$(\pi,0)$	**1**		*-4*	**4**		*4*		**-4**		**-1**				
$(\pi,\pi/3)$					4		-2				1	-1	-2	1
$(\pi,2\pi/3)$	**1**	-3	*2*	**1**		*-2*		**-1**	3	**-1**				
(π,π)					-8		-2				1	-4	4	4
$(2\pi/3,2\pi/3)$					2		2				1	-2	1	-2
$(\pi/3,\pi/3)$					-2		-2				1	2	1	-2

In this new matrix there are again 4 pairs of columns with similar elements permitting a substitution by an auxiliary. (Such a substitution makes sense whenever there are two columns c_1 and c_2 and two rows i and j so that $c_1[i]/c_2[i] = c_1[j]/c_2[j]$.) This time, however, the columns 0 and 9, 1 and 8, 2 and 5, and 3 and 7 do not only contain *some* "matching" coefficients, they are *entirely* suitable for substitution. Thus the matrix can be further optimized without any more auxiliaries by calculating

$$f_0 := f_0 - f_9 , \qquad f_1 := f_1 - f_8 ,$$
$$f_2 := f_2 - f_5 , \qquad f_3 := f_3 - f_7 .$$

The columns 5, 7, 8 and 9 are now totally empty and can be used to store further auxiliaries:

k	f_0	f_1	f_2	f_3	f_4	f_5	f_6	f_7	f_8	f_9	f_{10}	f_{11}	f_{12}	f_{13}
$(0,0)$					**8**		**2**				1	4	4	4
$(\pi/3,0)$	1	3	2	1										
$(2\pi/3,0)$					-4		2				1	1	-2	1
$(\pi,0)$	1		-4	4										
$(\pi,\pi/3)$					4		-2				1	-1	-2	1
$(\pi,2\pi/3)$	1	-3	2	1										
(π,π)					**-8**		**-2**				1	-4	4	4
$(2\pi/3,2\pi/3)$					*2*		*2*				1	**-2**	1	**-2**
$(\pi/3,\pi/3)$					*-2*		*-2*				1	2	1	-2

The bold, italic and underlined coefficients in above matrix show that there are still a lot of possible optimization steps to be done. Finally we might end up with the matrix

k	f_0	f_1	f_2	f_3	f_4	f_5	f_6	f_7	f_8	f_9	f_{10}	f_{11}	f_{12}	f_{13}
$(0,0)$					4		1			1				
$(\pi/3,0)$	1	1												
$(2\pi/3,0)$					1			-1			1			
$(\pi,0)$						1								
$(\pi,\pi/3)$								1			1			
$(\pi,2\pi/3)$	1	-1												
(π,π)							-1			1		-4		
$(2\pi/3,2\pi/3)$					2				1					
$(\pi/3,\pi/3)$									1			2		

produced by the following set of replacement rules

$$
\begin{array}{llllll}
f_{10} = f_0 + f_9 & , & f_{11} = f_1 + f_8 & , & f_{12} = f_2 + f_5 & , \\
f_{13} = f_3 + f_7 & , & f_0 = f_0 - f_9 & , & f_1 = f_1 - f_8 & , \\
f_2 = f_2 - f_5 & , & f_3 = f_3 - f_7 & , & f_5 = f_4 + f_6 & , \\
f_7 = 2f_4 - f_6 & , & f_6 = f_6 + 4f_4 & , & f_4 = f_{11} + f_{13} & , \\
f_{11} = f_{11} - f_{13} & , & f_8 = f_{10} + f_{12} & , & f_9 = f_{10} + 4f_{12} & , \\
f_{10} = f_{10} - 2f_{12} & , & f_5 = f_0 - 4f_2 & , & f_5 = f_5 + 4f_3 & , \\
f_0 = f_0 + f_3 & , & f_0 = f_0 + 2f_2 & , & f_1 = f_1 * 3 & , \\
f_6 = f_6 * 2 & , & f_7 = f_7 * 2 & . & &
\end{array}
$$

In summary we have found a way to calculate the 9 values $F(k)$ with only 31 additions and 14 multiplications.

B.3 The algorithm

How can the optimization of arbitrary given coefficient matrices be cast into an algorithmic form? The desired algorithm must consist of two parts: a first initializing part has to convert the given densely populated coefficient matrix into a sparse matrix (containing only a relatively small number of non-zero elements) and a set of replacement rules needed to perform the transition from the initial to the sparse matrix. Then in a second part the Fourier transform from $f(r)$ to $F(k)$ itself has to be performed using the optimized data calculated in part one.
Before dealing with the two parts in detail we first have to define the data structures for the "interface" between them, i.e. we have to ask in which form the sparse matrix and the replacement rules are stored in order to achieve maximum convenience and performance. Let us first of all introduce two notations: the number of r-points shall be called n, the number of k-points m with $m \leq n$. (Thus the original coefficient matrix has dimensions $m \times n$.) The matrix is accessed by rows during the Fourier transform in part two, hence a good choice is to store the non-zero matrix elements in row-major form as a one-dimensional (1D) array – we will call it cAr – of type

$$\text{Coefficient} = (i_{col}, value)\,,$$

where i_{col} is the element's column index and *value* its value. Additionally we need a 1D array of length m – called *rowEnd* – storing the position of the last coefficient of each row in cAr.
Regarding the replacement rules shown at the end of section B.2 we note that they all can be written as

$$f(i_{\text{target}}) = f(i_{\text{source1}}) + w\, f(i_{\text{source2}})\,,$$

where w is a real or complex weight coefficient. So a replacement rule can be stored in a data structure

$$\text{Rule} = (i_{\text{target}}, i_{\text{source1}}, i_{\text{source2}};\ w)$$

consisting of 3 integer and one real (or complex) data elements. The set of all replacement rules can then be stored in a simple 1D array of type "Rule".

```
(1)  FOR all columns c
         find best matching (c', n_match, ratio)
     N := max{n_match}
(2)  WHILE N > 1 {
         choose a pair (c, c') with n_match(c, c') = N
(3a)     IF c = ratio·c' {
             c -= ratio·c', c' := 0
             add rule (i_c, i_c, i_c', −ratio)
         }
(3b)     ELSE {
             c'' := first col. with no non-zero elem.
             FOR all matching elem. i of c and c' {
                 c''[i] := c[i] − ratio·c'[i]
                 c[i] := c'[i] := 0
             }
             add rule (i_c'', i_c, i_c', −ratio)
         }
(4)      FOR all columns c''
             IF c'' or its best match equals c or c'
                 find new (col, n_match, ratio) for c''
         N := max{n_match}
(5)      find all col. {c} containing only 1 elem.≠ 0
         WHILE two of these c have their elem.≠ 0
                  in the same row i {
             c1[i] += c2[i] , c2[i] := 0
         }
     }
(6)  store remaining non-zero matrix elements...
     ... in sparse matrix format
```

Figure B.2: SFFT initialization algorithm: create optimized (sparse) Fourier coefficient matrix and replacing rules for original values $f(\mathbf{r})$ and auxiliaries.

```
(1) FOR all replacement rules r
        f(r.i_target) := f(r.i_source1) + r.w · f(r.i_source2)
(2) F(k) := (0, ..., 0)
    coefNr := 0
    FOR all k's
        WHILE coefNr ≤ rowEnd(k) {
            F(k) += f(cAr[coefNr].i_col)
                        ·cAr[coefNr].value
            coefNr += 1
        }
```

Figure B.3: SFFT: the Fourier transform itself, using the previously created replacement rules and sparse matrix.

After these preliminaries we come to the initialization part of SFFT. Figure B.2 shows the basic steps of such an algorithm in a "pseudo-code" language (see figure B.2). The first major task of this algorithm is to identify those couples of matrix columns $(\mathbf{c}, \mathbf{c}')$ that "match best" in the sense that combining these columns into an auxiliary allows maximum benefit in form of as many non-zero elements of $\mathbf{c}$ and $\mathbf{c}'$ being canceled out as possible (part (1) of figure B.2). Then we enter an "auxiliary-introducing" loop that continues until there is no more benefit achievable from introducing new auxiliaries ($N = 1$) (part (2)). If two columns $(\mathbf{c}, \mathbf{c}')$ with best optimization gain are identified some unused storage has to be found where the new auxiliary can be deposited. If the replacement affects *all* elements of $\mathbf{c}$ the old data in $\mathbf{c}$ can be overwritten by the auxiliary (part (3a)). Otherwise the auxiliary is stored in the first completely empty column of the matrix (part (3b)). In part (4) the "best-matching" information of all column tuples containing $\mathbf{c}$ or $\mathbf{c}'$ has to be updated since $\mathbf{c}$ and $\mathbf{c}'$ have been modified. Part (5) is some "garbage collection": if there are several columns containing only one non-vanishing matrix element in the same row, these columns can be added up and the result can be stored in one of them, which frees storage for new auxiliaries. Finally part (6) transforms the remaining, almost empty coefficient matrix into the sparse storage format discussed above.

What does this initialization "cost"? Part (1) has complexity $O(n^2m)$, the loop in part (2) is $O(n \ln m)$ and its internal loops all need $O(n)$ or $O(m)$ operations. Altogether we end up with a complexity of $O(n^2m)$. This means that even for huge input data sets of $n = 10^5$ variables and if relatively large subsets of $F(k)$ are wanted ($m \approx 10^2...10^3$) the initialization process finishes reasonably fast on modern high-performance computers. The optimized coefficient array can be used for millions of Fourier transforms once it has been determined, hence the extra initialization work is a good "investment" in many cases.

Compared to the algorithm described above the optimized Fourier transform itself is remarkably simple (figure B.3). Part (1) applies the replacement rules onto the initial values $f(r)$ (and onto the empty prolongation of array $f(r)$ destinated to store supplementary auxiliaries). The initial values $f(r)$ are thereby overwritten. Part (2) then calculates the sparse matrix-vector product of these modified $f(r)$ values with the optimized Fourier coefficient matrix.

B.4 Implementation in C++

An object oriented C++ implementation of the algorithms described above will be available from the CPC program library. The code consists of the files

sfft.h : user interface definition, documentation
sfft.ih : implementation of template classes
sfft_.cc : implementation of non-template classes

and can be used to optimize matrices of *real* Fourier coefficients (see Fig. B.4). Figure B.4 shows the main features of the program package: class Real_SFFT<.,.> provides the whole functionality. The two template parameters specify the user's prefered 1D array type for element types double and int. Default values are valarray<double> and valarray<int>. An arbitrary real coefficient matrix can be assigned and optimized via the initialize() member function; member function nAuxiliaries() returns the number of empty array elements which have to be appended to the array $f(r)$ of input data and which are needed to store auxiliaries. The Fourier transform itself is performed by perform_FT() if $f(r)$ is scalar valued and by perform_multiple_FTs() if $f(r) = \boldsymbol{f}(r)$ is a vector valued function.

B.5 Summary

The "SFFT" algorithms and program codes described in this paper serve to speed up "selective" discrete Fourier transformations in which a relatively small set of transformed values $\{F(k)\}$ is to be calculated from a vast set $\{f(r)\}$ of original data points. The algorithms can be described as "hand-taylored" variants of standard FFT codes: both methods try to avoid multiple calculation of almost identical parts of the whole matrix-vector multiplication transforming $\{f(r)\}$ into $\{F(k)\}$. To this purpose FFT routines use a fix recursive "divide and conquer" scheme while SFFT thoroughfully analyses the Fourier coefficient matrix and generates a series of replacement rules and auxiliaries that eliminate avoidable calculations in an almost optimal way. The price to be paid for this increased flexibility and optimization is that SFFT requires a non-trivial initialization procedure in which the matrix is analysed. SFFT performes best if the number of transformed data points $|\{F(k)\}|$ is significantly smaller than $|\{f(r)\}|$ and significantly bigger than 1 and

- if a large number of Fourier transforms with varying input data $f(r)$ but constant Fourier coefficients has to be performed and/or
- if the functions $\boldsymbol{f}(r)$ and $\boldsymbol{F}(k)$ are vector valued with large vector dimensions.

In these cases the codes are often by more than one order of magnitude faster than the best standard FFT methods.

```
#include ßfft.h"
#include <valarray.h>

typedef valarray<double> dArray;
typedef valarray<int> iArray;

const double COEFS[90] =
{ 1, 4, 4, 4, 8, 4, 2, 4, 4, 1,
  1, 3, 2, 1, 0,-2, 0,-1,-3,-1,
  1, 1,-2, 1,-4,-2, 2, 1, 1, 1,
  1, 0,-4, 4, 0, 4, 0,-4, 0,-1,
  1,-1,-2, 1, 4,-2,-2, 1,-1, 1,
  1,-3, 2, 1, 0,-2, 0,-1, 3,-1,
  1,-4, 4, 4,-8, 4,-2, 4,-4, 1,
  1,-2, 1,-2, 2, 1, 2,-2,-2, 1,
  1, 2, 1,-2,-2, 1,-2,-2, 2, 1 };
const double F_R[10] =
{ 1., 0., 2., 1., 1., 3., 0., 1., 0., 2.  };

int main()
{
  dArray coefs( COEFS, 90 )
  SFFT::Real_SFFT< dArray, iArray > sfft;
  sfft.initialize( coefs, 10, 9 );

  dArray f_r( 10+sfft.nAuxiliaries() );
  dArray F_k( 9 );
  for( int i=0; i<10; i++ )
    f_r[i] = F_R[i];

  sfft.perform_FT( f_r, F_k );
  for( int j=0; j<9; j++ )
    cout << F_k[j] << endl;
}
```

Figure B.4: Example Program showing the use of the C++ SFFT code package

Bibliography

[1] E. Dagotto, Rev. Mod. Phys. **66**, 763 (1994).

[2] R. Eder, *Numerical studies of strongly correlated electrons*, Habilitationsschrift, Würzburg (1998).

[3] J.G. Bednorz and K.A. Müller, Z. Phys. B **64**, 189 (1986).

[4] M. Imada, A. Fujimori, Y. Tokura, Rev. Mod. Phys. **70**, 1039 (1998).

[5] N.W. Ashcroft und N.D. Mermin, *Solid state physics*, Saunders College, Philadelphia (1981).

[6] C. Kittel, *Introduction to Solid State Physics*, John Wiley and Sons, New York, 7th ed. (1995).

[7] W. Hanke, A. Muramatsu, G. Dopf, Phys. Blätt. **47**, 1061 (1991).

[8] S. Wolfram *The MATHEMATICA Book*, Cambridge University Press, 3rd ed. (1999).

[9] H. Eck, "FeynArts 2.0 – Development of a generic Feynman diagram generator", PhD thesis, Universität Würzburg (1995); FeynArts homepage at `http://www.feynarts.de`.

[10] W. Nolting, *Grundkurs Theoretische Physik*, vol. 7, Neufang, Ulmen, 2. Aufl. (1992).

[11] J. Hubbard, Proc. Roy. Soc. A **276**, 238 (1963).

[12] M.G. Zacher, A. Dorneich, C. Gröber, R. Eder, and W. Hanke, *The Metal-Insulator Transition in the Hubbard Model*, in *High Performance Computing in Science and Engineering '99*, Springer Verlag (2000).

[13] G. Schütz *et al.*, Phys. Rev. Lett. **58**, 737 (1987); C.T. Chen *et al.*, Phys. Rev. B **42**, 7262 (1990); T. Koide *et al.*, Phys. Rev. B **44**, 4697 (1991), and Jpn. J. Appl. Phys. **32**, suppl. 32-2 (1993).

[14] A. Dorneich, R. Eder, E. Goehring, to be published.

[15] A. Dorneich, M.G. Zacher, C. Gröber, R. Eder, Phys. Rev. B **61**, 12816 (2000).

[16] A. W. Sandvik, Phys. Rev. B **59**, R14157 (1999).

[17] A. W. Sandvik, Phys. Rev. B **56**, 11678 (1997).

[18] W. von der Linden, Phys. Rep. **220**, 53 (1992).

[19] H. G. Evertz, G. Lana and M. Marcu, Phys. Rev. Lett. **70**, 875 (1993).

[20] A. Dorneich and M. Troyer, submitted to Phys. Rev. E, cond-mat/0106471.

[21] K. Bernardet, G.G. Batrouni, M. Troyer, A. Dorneich, *Destruction of Superfluid and Long Range Order by Impurities in Two Dimensional Systems*, to be published in "High Performance Computing in Science and Engineering 2001" (Springer Verlag).

[22] F. Hebert, G.G. Batrouni, R.T. Scalettar, G. Schmid, M. Troyer, A. Dorneich, submitted to Phys. Rev. B, cond-mat/0105450.

[23] S.C. Zhang, J.P. Hu *et al.*, Phys. Rev. B **60**, 13070 (1999).

[24] J. Hubbard, Proc. Roy. Soc. A **277**, 237 (1964); J. Hubbard, Proc. Roy. Soc. A **281**, 401 (1964).

[25] C. Gröber, M. G. Zacher, and R. Eder, cond-mat/9902015.

[26] S.C. Zhang, Science **275**, 1089 (1997).

[27] A. Dorneich, W. Hanke, E. Arrigoni, M. Troyer, and S.C. Zhang, cond-mat/0106473, submitted to Phys. Rev. Lett.

[28] A. Dorneich, W. Hanke, E. Arrigoni, M. Troyer, and S.C. Zhang, to be published in the Proceedings of the ISSP Symposium on "Correlated Electrons", Tokyo (2001).

[29] A. Dorneich, W. Hanke, E. Arrigoni, M. Troyer and S.C. Zhang, will be submitted to Phys. Rev. B.

[30] Xiao Hu, Phys. Rev. Lett. **87**, 057004 (2001).

[31] A. Dorneich, *A fast algorithm for selective Fourier transformations*, submitted to Comp. Phys. Comm.

[32] D. van der Marel and G.A. Sawatzky, Phys. Rev. B **37**, 10674 (1988).

[33] F.C. Zhang und T.M. Rice, Phys. Rev. B **37**, 3759 (1988).

[34] see, for example, S.R. White, D.J. Scalapino, R.L. Sugar, E.Y. Loh, J.E. Gubernatis, R.T. Scalettar, Phys. Rev. B **40**, 506 (1989).

[35] F. Gebhard, *The Mott Metal-Insulator Transition*, Springer, Berlin/Heidelberg/New York (1997).

[36] A. Brooks Harris and R. V. Lange, Phys. Rev. **157**, 295 (1967).

[37] P.W. Anderson, Science **235**, 1196 (1987).

[38] M.S. Hybertsen, E.B. Stechel, M. Schlüter, and D.R. Jennison, Phys. Rev. B **41**, 11068 (1990).

[39] L.F. Feiner, J.H. Jefferson, and R. Raimondi, Phys. Rev. B **53**, 8751 (1996); R. Raimondi, J.H. Jefferson, and R. Raimondi, Phys. Rev. B **53**, 8774 (1996);

[40] V.I. Belinicher and A.L. Chernyshev, Phys. Rev. B **49**, 9746 (1994); V.I. Belinicher, A.L. Chernyshev, and L.V. Popovich, Phys. Rev. B **50**, 13768 (1994);

[41] M. Ogata, M.U. Luchini, and T.M. Rice, Phys. Rev. B. **44**, 12083 (1991).

[42] T.M. Rice, in *The Physics and Chemistry of Oxide Superconductors*, edited by Y. Tye and H. Yasuoka, Springer Verlag, Berlin (1992).

[43] C. Herring, *Direct Exchange Between Well Separated Atoms*, in Magnetism, vol. 2B, edited by G.T. Rado and H. Suhl, Academic Press, New York (1965).

[44] S. Sachdev and R.N. Bhatt, Phys. Rev. B, **41**, 9323 (1990).

[45] S. Gopalan, T.M. Rice and M. Sigrist, Phys. Rev. B **49**, 8901 (1994).

[46] E. Altman and A. Auerbach, cond-mat/0108087 (unpublished).

[47] M.N. Barber, *Finite-size scaling*, in *Phase Transitions*, vol. 8, Academic Press, London (1983).

[48] C. Lanczos, J. Res. Nat. Bur. Stand. **45**, 255 (1950).

[49] S. R. White, Phys. Rev. Lett. **69**, 2863 (1992).

[50] S. R. White, Phys. Rev. B **48**, 10345 (1993).

[51] R. Noack, S. R. White, D. J. Scalapino, Phys. Rev. Lett. **73**, 886 (1994).

[52] M. Suzuki, Prog. Theor. Phys. **65**, 1454 (1976).

[53] J. E. Hirsch, R. L. Sugar, D. J. Scalapino and R. Blankenbecler, Phys. Rev. B **26**, 5033 (1982).

[54] J. E. Hirsch, Phys. Rev. B **31**, 4403 (1985).

[55] J.E. Hirsch, Phys. Rev. B **38**, 12023 (1988).

[56] E. Loh and J. Gubernatis, *Electronic Phase Transitions*, ed. W. Hanke and Y. V. Kopaev, North Holland, Amsterdam (1992).

[57] R. Preuss, W. Hanke and W. von der Linden, Phys. Rev. Lett. **75**, 1344 (1995).

[58] R. Preuss, W. Hanke *et al*, Phys. Rev. Lett. **79**, 1122 (1997).

[59] B. Brendel, *Numerical and analytical analysis of low-dimensional strongly correlated electron systems*, PhD thesis, Universität Würzburg (1999).
(available from `ftp://ftp.physik.uni-wuerzburg.de/pub/dissertation/`).

[60] W. Kinzel and G. Reents, *Physik per Computer*, Spektrum Verlag, Heidelberb, Berlin, Oxford (1996).

[61] R.H. Landau and M.J. Páez, *Computational Physics*, John Wiley & Sons, New York (1997).

[62] M. Metropolis, A.W. Rosenbluth, A.H. Teller, and E. Teller, J. Phys. Chem. **21**, 1087 (1953).

[63] A.P. Harju, E. Arrigoni, W. Hanke, B. Brendel, and S. Kivelson, submitted to Phy. Rev. Lett., cond-mat/0012293 v5 (2001).

[64] H.Q. Lin and J.E. Gubernatis, Computers in Physics **7**, 400 (1993).

[65] A. Dorneich, *Exakte-Diagonalisierungs-Studien zur SO(5)-Symmetrie an Leitermodellen*, Diplomarbeit, Universität Würzburg (1998).
(available from `ftp.physik.uni-wuerzburg.de/pub/diplom/dorneich.ps.gz`)

[66] J. Jaklič and P. Prelovšek, Phys. Rev. B **49**, 5065 (1994).

[67] A.L. Fetter and J.D. Walecka, *Quantum Theory of Many-Particle Systems*, Mc Graw-Hill, New York (1971).

[68] S. Doniach and E.H. Sondheimer, *Green's functions for solid state physicists*, Benjamin/Cummings, Reading, Mass. (1982).

[69] J. E. Gubernatis, M. Jarrell, R. N. Silver and D. S. Silvia, Phys. Rev. B **44**, 6011 (1991).

[70] W. von der Linden, R. Preuss and W. Hanke, J. Phys.: Condens. Matter **8**, 3881 (1996).

[71] J.K. Cullum and R.A. Willoughby, *Lanczos Algorithms for Large Symmetric Eigenvalue Computations*, Birkhäuser Verlag, Boston (1985).

[72] J. Stoer, *Numerische Mathematik 1*, Springer, Berlin-Heidelberg-New York, 6. Auflage (1994),

[73] J. Stoer and R. Bulirsch, *Numerische Mathematik 2*, Springer, Berlin-Heidelberg-New York, 2. Auflage (1990).

[74] S. Kaniel, Math. Comp. **20**, 369 (1966).

[75] C. C. Paige, *The computation of eigenvalues and eigenvectors of very large sparse matrices*, PhD thesis, London University (1971).

[76] Y. Saad, SIAM J. Num. Anal. **17**, 687 (1980).

[77] P. Fulde, *Electron correlations in molecules and solids*, Springer, Berlin-Heidelberg-New York, 3. erw. Aufl. 1995.

[78] S.W. Haas, *Dynamical properties of strongly correlated fermionic systems*, PhD thesis, Florida State University, 1995.

[79] G. Fano, F. Ortolami, and F. Semeria, Int. J. Mod. Phys. B **3**, 1845 (1990).

[80] G. Fano, F. Ortolami, and A. Parola, Phys. Rev. B **46**, 1048 (1992).

[81] H.Q. Lin, Phys. Rev. B **42**, 6561 (1990).

[82] M. Troyer and A. Läuchli, private communication.

[83] B.H. Bransden and C.J. Joachain, *The Physics of Atoms and Molecules*, Longman (1982).

[84] J.C. Slater, *Quantum Theory of Atomic Structure*, vol. I and II, McGraw-Hill, New York (1960).

[85] D. Vollhardt in "Correlated Electron Systems, Vol. 9, ed. by V. J. Emery, World Scientific, Singapore (1993).

[86] W. Metzner, D. Vollhardt, Phys. Rev. Lett. **62**, 324 (1989).

[87] E. Müller-Hartmann, Z. Phys. B **74**, 507 (1989).

[88] W. Metzner, Z. Phys. B **77**, 253 (1989).

[89] A. Bayer, *Zirkulardichroische Untersuchungen und Momentenanalyse an epitaktisch gewachsenen CrO_2-Filmen*, Diplomarbeit, Universität Würzburg (2001).

[90] B.T. Thole, P. Carra, F. Sette, and G. van der Laan, Phys. Rev. Lett. **68**, 1943 (1992).

[91] P. Carra, B.T. Thole, M. Altarelli, and X. Wang, Phys. Rev. Lett. **70**, 694 (1993).

[92] H.G. Evertz and M. Marcu, in *Lattice 92*, Amsterdam 1992, ed. J. Smit et al., Nucl. Phys. B (Proc. Suppl.) **30**, 277 (1993), and in *"Quantum Monte Carlo Methods in Condensed Matter Physics"*, ed. M. Suzuki, World Scientific (1994), p. 65.

[93] U.-J. Wiese and H.-P. Ying, Phys. Lett. A **168**, 143 (1992); Z. Phys. B **93**, 147 (1994).

[94] N. Kawashima, J.E. Gubernatis, and H.G. Evertz, Phys. Rev. B **50**, 136 (1994).

[95] N. Kawashima and J.E. Gubernatis, Phys. Rev. Lett. **73**, 1295 (1994).

[96] N. Kawashima, J. Stat. Phys. **82**, 131 (1996).

[97] N. Kawashima and J.E. Gubernatis, J. Stat. Phys. **80**, 169 (1995).

[98] H.G. Evertz, in *"Numerical Methods for Lattice Quantum Many-Body Problems"*, ed. D.J. Scalapino, Addison Wesley Longman, Frontiers in Physics (1998).

[99] M. Kohno and M. Takahashi, Phys. Rev. B **56**, 3212 (1997).

[100] A. W. Sandvik and J. Kurkijärvi, Phys. Rev. B **43**, 5950 (1991); A. W. Sandvik, J. Phys. A **25**, 3667 (1992).

[101] M. Takasu, S. Miyashita and M. Suzuki, Prog. Theor. Phys. **75**, 1254 (1986).

[102] N. Hatano, J. Phys. Soc. Japan **63**, 1691 (1993).

[103] F. Assaad, W. Hanke and D.J. Scalapino, J. Phys. Rev. B **50**, 12385 (1994).

[104] B. Li, N. Madras and A. D. Sokal, J. Statist. Phys. **80**, 661-754 (1995).

[105] A. W. Sandvik, R. R. P. Singh and D. K. Campbell, Phys. Rev. B **56**, 14510 (1997).

[106] A detailed description of the new method will soon be submitted to Comp. Phys. Comm., and a C++ code package will be available from the CPC program library.

[107] W. von der Linden,R. Preuss and W. Hanke, J. Phys.: Condens. Matter **8**, 3881 (1996).

[108] M. Jöstingmeier, *Numerische Studien zur bosonischen Beschreibung von Spinsystemen*, Diplomarbeit, Universtät Würzburg, 2000.

[109] G.T. Zimanyi, P.A. Crowell, R.T. Scalettar, G.G. Batrouni, Phys. Rev. B **50**, 6515 (1994).

[110] M. Cha, M.P.A. Fisher, S.M. Girvin, M. Wallin, and A.P. Young, Phys. Rev. B **44**, 6883 (1991).

[111] E.S. Sorensen, M. Wallin, S.M. Girvin, and A.P. Young, Phys. Rev. Lett. **69**, 828 (1992).

[112] G.G. Batrouni, B. Larson, R.T. Scalettar, J. Tobochnik, and J. Wang, Phys. Rev. B **48**, 9628 (1993).

[113] K.J. Runge, Phys. Rev. B **45**, 13136 (1992).

[114] V.F. Elesin, V.A. Kashurnikov, and L.A. Openov, JETP Lett. **60**, 177 (1994).

[115] A.P. Kampf and G.T. Zimanyi, Phys. Rev. B **47**, 279 (1993).

[116] R.T. Scalettar, G.G. Batrouni, and G.T. Zimanyi, Phys. Rev. Lett. **66**, 3144 (1991).

[117] G.T. Zimanyi, P.A. Crowell, R.T. Scalettar, and G.G. Batrouni, Phys. Rev. B **50**, 6515 (1994).

[118] K.G. Singh and D.S. Rokhsar, Phys. Rev. B **46**, 3002 (1992).

[119] M. Kohno and M. Takahashi, Phys. Rev. B **56**, 3212 (1997).

[120] G.G. Batrouni, R.T. Scalettar, G.T. Zimanyi, and A.P. Kampf, Phys. Rev. Lett. **74**, 2527 (1995); R.T. Scalettar, G.G. Batrouni, A.P. Kampf, and G.T. Zimanyi, Phys. Rev. B **51** (1995) 8467.

[121] G.G. Batrouni and R.T. Scalettar, Phys. Rev. Lett. **84** 1599 (2000).

[122] E. Dagotto and T.M. Rice, Science **271**, 618 (1996).

[123] Y. Iino and M. Imada, J. Phys. Soc. Jpn. **65**, 3728 (1996).

[124] R. Eder, unpublished.

[125] P.O. Sushkov, Phys. Rev. B **62**, 12135 (2000).

[126] M. Troyer, H. Tsunetsugu, and D. Würtz, Phys. Rev. B **50**, 13515 (1994).

[127] L.M. Roth, Phys. Rev. **184**, 451 (1969).

[128] G. Geipel and W. Nolting, Phys. Rev. B **38**, 2608 (1988); W. Nolting and W. Borgiel, Phys. Rev. B **39**, 6962 (1989).

[129] B. Mehlig, H. Eskes, R. Hayn, and M.B.J. Meinders, Phys. Rev. B **52**, 2463 (1995).

[130] J. Beenen and D.M. Edwards, Phys. Rev. B **52**, 13636 (1995).

[131] R. Micnas, J. Ranninger, and S. Robaszkiewcz, Rev. Mod. Phys. **62**, 113 (1990).

[132] A. Avella, F. Mancini, D. Villani, L. Siurakshina, V.Yu. Yushankhai, Int. J. Mod. Phys. B **12**, 81 (1998); see also F. Mancini, Phys. Lett. A **249**, 231 (1998).

[133] C.N. Yang, Phys. Rev. Lett. **63**, 2144 (1989); S.C. Zhang, Phys. Rev. Lett. **65**, 120 (1990).

[134] S. Pairault, D. Senechal, and A.M.S. Tremblay, Phys. Rev. Lett. **80**, 5389 (1998).

[135] R. Eder, O. Rogojanu, and G.A. Sawatzky, Phys. Rev. B **58**, 7599 (1998).

[136] C. Gröber, R. Eder, W. Hanke (in preparation).

[137] L.N. Bulaevskii, E.L. Nagaev, and D.I. Khomskii, Sov. Phys. JETP **27**, 638 (1967).

[138] S.A. Trugman, Phys. Rev. B **37**, 1597 (1988); ibid. **41**, 892 (1990).

[139] R. Eder and K.W. Becker, Z. Phys. B **78**, 219 (1990).

[140] O.P. Sushkov, Solid State Commun. **83**, 303 (1992).

[141] G. Reiter, Phys. Rev. B **49**, 1536 (1994).

[142] R. Hayn, A.F. Barabanov, J. Schulenburg, and J. Richter, Phys. Rev. B **53**, 11714 (1996).

[143] M.G. Zacher, *From one to two dimensions: Numerical and analytical studies of strongly correlated electron systems*, PhD thesis, Universität Würzburg (1999). (available from `ftp://ftp.physik.uni-wuerzburg.de/pub/dissertation/`).

[144] E. Arrigoni and W. Hanke, Phys. Rev. Lett. **82**, 2115 (1999).

[145] J. Bardeen, L.N. Cooper, J.R. Schrieffer, Phys. Rev. **108**, 1175 (1957).

[146] G. Burns, *High-Temperature Superconductivity*, Academic, New York, 1992.

[147] E. Demler and S.C. Zhang, Phys. Rev. Lett **75**, 4126 (1995).

[148] R. Eder, W. Hanke and S.C. Zhang, Phys. Rev. B **57**, 13781 (1998).

[149] R. Eder, A. Dorneich, M.G. Zacher, W. Hanke and S.C. Zhang, Phys. Rev. B **59**, 561 (1999).

[150] M. Calandra and S. Sorella, Phys. Rev. B **61**, R11894 (2000).

[151] V.J. Emery, S. Kivelson and H.Q. Lin, Phys. Rev. Lett. **64**, 475 (1990).

[152] M. Salmhofer, Commun. Math. Phys. **194**, 249 (1998).

[153] C.J. Halboth and W. Metzner, Phys. Rev. B **61**, 7364 (2000).

[154] C. Honerkamp, M. Salmhofer, N. Furukawa, and T.M. Rice, Phys. Rev. B **63**, 035109 (2001).

[155] H. Georgi, *Lie algebras in particle physics*, Benjamin/Cummings, Reading, Mass., 1982.

[156] W. Hanke, R. Eder, E. Arrigoni, A. Dorneich, S. Meixner and M.G. Zacher, *SO(5) Symmetry in t-J and Hubbard Models*, *Festkörperprobleme/ Advances in Solid State Physics* vol. 38, Vieweg Verlag (1999).

[157] W. Heisenberg, Z. Physik **77** (1932).

[158] C.L. Henley, unpublished manuscript.

[159] J.P. Hu and S.C. Zhang, *Review of SO(5) Group Theory*, to be published.

[160] S. Meixner, W. Hanke, E. Demler and S.C. Zhang, Phys. Rev. Lett. **79**, 4902 (1997).

[161] S. Rabello, H. Kohno, E. Demler and S.C. Zhang, Phys. Rev. Lett. **80**, 3586 (1998).

[162] H.H. Lin, L. Balents and M.P.A. Fisher, Phys. Rev. B **58**, 1794 (1998).

[163] J.R. Schrieffer, X.G. Wen and S.C. Zhang, Phys. Rev. B **39**, 11663 (1989).

[164] C. Burgess and A. Lutken, Phys. Rev. B **57**, 8642 (1998).

[165] X. Hu, T. Koyama and M. Tachiki, Phys. Rev. Lett. **82**, 2568 (1999).

[166] E. Demler, H. Kohno, and S.-C. Zhang, Phys. Rev. B **58**, 5719 (1998).

[167] Y. Bazaliy, E. Demler, and S.-C. Zhang, Phys. Rev. Lett. **79**, 1921 (1997).

[168] D. Arovas, A.J. Berlinsky, C. Kallin, and S.C. Zhang, Phys. Rev. Lett. **79**, 2871 (1997).

[169] E. Demler, A.J. Berlinsky, C. Kallin, G. Arnold, and M. Beasley, Phys. Rev. Lett. **80**, 2917 (1998).

[170] D. Sheehy and P. Goldbart, Phys. Rev. B **57**, 8131 (1998).

[171] P.M. Goldbart and D. Sheehy, Phys. Rev. B **58**, 5731 (1998).

[172] M. Greiter, Phys. Rev. Lett. **79**, 4898 (1997).

[173] G. Baskaran and P.W. Anderson, J. Phys. and Chem. of Solids **59**, 1780 (1998).

[174] S.C. Zhang, J. Phys. and Chem. of Solids **59**, 1774 (1998).

[175] F. Ronning *et al.*, Science **282**, 2067 (1998).

[176] M. G. Zacher, W. Hanke, E. Arrigoni, and S.C. Zhang, Phys. Rev. Lett. **85**, 824 (2000).

[177] E. Arrigoni, M. G. Zacher, and W. Hanke, cond-mat/0105125 (unpublished).

[178] E. Arrigoni and W. Hanke, Phys. Rev. Lett. **82**, 2115 (1999).

[179] J. M. Kosterlitz, D. R. Nelson and M. E. Fisher, Phys. Rev. B **13**, 412 (1976).

[180] A. Aharony, cond-mat/0107585 (unpublished).

[181] A. Ino, T. Mizokawa and A. Fujimori, Phys. Rev. Lett. **79**, 2101 (1998).

[182] B.O. Wells, Y.S. Lee *et al.*, Science **277**, 1067 (1997).

[183] V.L. Berezinskii, Zh. Éksp. Teor. Fiz. **59**, 907 (1970) [Sov. Phys. JETP **32**, 493 (1971)].

[184] J.M. Kosterlitz and D.J. Thouless, J. Phys. C **6**, 1181 (1973).

[185] H.Q. Ding, Phys. Rev. B **45**, 230 (1992).

[186] D. R. Nelson and J. M. Kosterlitz, Phys. Rev. Lett. **39**, 1201 (1977).

[187] E.L. Pollock and D.M. Ceperly, Phys. Rev. B **36**, 8343 (1987).

[188] K. Harada and N. Kawashima, Phys. Rev. B **55**, R11949 (1997).

[189] M. Troyer and S. Sachdev, Phys. Rev. Lett. **81**, 5418 (1998).

[190] S. Chakravarty, B.I. Halpertin and D.R. Nelson, Phys. Rev. Lett. **60**, 1057 (1988).

[191] S. Chakravarty, B.I. Halpertin and D.R. Nelson, Phys. Rev. B **39**, 2344 (1989).

[192] H.F. Fong, P. Bourges, Y. Sidis, L.P. Regnault, J. Bossy, A. Ivanov, D.L. Milius, I.A. Aksay, B. Keimer, Phys. Rev. B **61**, 14773 (2000).

[193] H. He, Y. Sidis, P. Bourges, G.D. Gu, A. Ivanov, N. Koshizuka, B. Liang, C.T. Lin, L.P. Regnault, E. Schoenherr, and B. Keimer, Phys. Rev. Lett. **86**, 1610 (2001).

[194] P. Pfeuty *et al.*, Phys. Rev. B **10**, 2088 (1974).

[195] D. Blankschtein and A. Aharony, Phys. Rev. Lett. **47**, 439 (1981).

[196] J.P. Hu and S.C. Zhang, cond-mat/0005334 (invited talk at the M2S conference in Houston).

[197] H. Takagiwa, J. Akimitsu *et al.*, J. Phys. Soc. Jpn. **70**, 333 (2001).

[198] S. Mitsuda *et al.*, J. Phys. Soc. Jpn **61**, 4667 (1992).

[199] Z.Q. Peng, K. Krug and K. Winzer, Phys. Rev. B **57**, 8123 (1998).

[200] H. Kobayashi, A. Kobayashi and P. Cassoux, Chem. Soc. Rev. **29**, 325 (2000).

[201] M. Troyer *et al.*, Lecture Notes in Computer Science 1505, 191 (1998).

[202] Further information available from `http://www.mpi-forum.org`.

[203] J. W. Cooley and J. W. Tukey, Math. Comput. **19**, 297-301 (1965).

[204] FFTW documentation and code can be obtained from `http://www.fftw.org`.

Zusammenfassung

Die vorliegende Arbeit umfasst zwei große Teile. In einem ersten, "technischen" Teil wurden wichtige aktuelle Verfahren zur numerischen Simulation hoch korrelierter Elektronensysteme vorgestellt; für drei dieser Verfahren wurden anschließend eigene Beiträge zur Weiterentwicklung und Verbesserung beschrieben: für die Exakte Diagonalisierung (ED), die Quanten-Monte-Carlo-Technik der Stochastischen Reihenentwicklung (SSE) sowie computerunterstützte algebraische Termumformungen (computer aided algebraic manipulations, CAAM). Mehrere konkrete Implementierungen wurden in ihren Grundgedanken erläutert; die Leistungsfähigkeit jedes dieser Computerprogramme wurde anschließend in exemplarischen Anwendungen auf moderne Forschungsprobleme nachgewiesen. In einem zweiten, "physikalischen" Teil wurden neue numerische Ergebnisse zur SO(5)-Theorie der Hochtemperatur-Supraleitung zusammengefasst dargestellt.
Der erste Teil der Arbeit stellte zunächst die wichtigsten mikroskopischen Modelle und Simulationstechniken für Vielelektronensysteme vor. Dabei wurde gezeigt, wie aus mikroskopischen fermionischen Modellen effektive Modelle konstruiert werden können, in der die niederenergetische Physik des Fermionensystems durch bosonische Quasiteilchen ausgedrückt wird. Insbesondere wurde ein Vier-Bosonen-Modell aus drei Triplett-Bosonen und einem Lochpaar-Boson als ein effektives Modell für zweidimensionale Elektronensysteme hergeleitet, die sich auf mikroskopischer Ebene durch das Heisenberg- oder T-J-Modell beschreiben lassen. Das wichtigste reale Anwendungsgebiet dieser Modelle sind Kuprat-Supraleiter mit ihren zweidimensionalen Kupferdioxidebenen.
Kapitel 2 war der Simulationstechnik der Exakten Diagonalisierung (ED) und ihren Anwendungsmöglichkeiten gewidmet. Nach einer Beschreibung der Stärken und Beschränkungen dieser Methode wurden einige Ansätze und Tricks für eine effiziente Implementierung auf modernen Höchstleistungsrechnern vorgestellt. Zwei besonders wichtige Einsatzfelder wurden identifiziert: die Simulation niedrigdimensionaler regelmäßiger Kristallgitter – Atomketten, Leitern und Ebenen – mit periodischen Randbedingungen sowie die Berechnung einzelner Einheitszellen komplexer Metall-Nichtmetall-Verbindungen. Ein Beispiel für die erste Anwendungsklasse sind die Kuprat-Supraleiter, in die zweite Klasse fallen etwa die Metall-Liganden-Cluster magnetischer $3d$-Metalloxide. Am Ende des Kapitels wurden zwei aktuelle Probleme der theoretischen Festkörperphysik mit Hilfe der Exakten Diagonalisierung untersucht. Eine Studie der Metall-Isolator-Übergänge im Hubbard-Modell

bei Halbfüllung – ein weithin akzeptiertes mikroskopisches Modell für die undotierten Ausgangsmaterialien der Kuprat-Supraleiter – enthüllte ein vielgestaltiges Phasendiagramm: neben zwei verschiedenen metallischen Phasen, existiert eine antiferromagnetisch geordnete isolierende Phase (Mott-Heisenberg-Isolator) und eine weitere nichtleitende Phase, in der das isolierende Verhalten ausschließlich von elektrostatischen Wechselwirkungen zwischen den Elektronen verursacht wird (Mott-Hubbard-Isolator). In einer zweiten exemplarischen Anwendung der ED wurden experimentell gemessene magnetische Röntgen-Zirkulardichroismus-Spektren des technologisch interessanten magnetischen Halbmetalls CrO_2 im Computerexperiment reproduziert. Dabei wurde ein einzelner oktaedrischer Cluster aus einem zentralen Chromatom und sechs umgebenden Sauerstoffatomen untersucht. Der Vergleich zwischen Experiment und numerischer Simulation, der mit großer Genauigkeit gelang, ermöglicht es, die wichtigsten elementaren Wechselwirkungsprozesse in dem Material zu bestimmen. Die Werte der mikroskopischen Wechselwirkungsparameter – Kristallfeldstärken, Coulomb- und Austauschintegrale und Spin-Bahn-Kopplungen – konnten auf diese Weise sogar quantitativ bestimmt werden.

In Kapitel 3 wurde die Quanten-Monte-Carlo-Simulationstechnik der stochastischen Reihenentwicklung (SSE) vorgestellt und anschließend eine Erweiterung dieser Technik entwickelt: eine Methode zur effizienten Messung beliebiger Greenscher Funktionen. In Benchmark-Tests an Heisenberg-Modellen in ein, zwei und drei Dimensionen wurde gezeigt, dass der Rechenaufwand als Funktion der Systemgröße und der inversen Temperatur in der SSE-Methode annähernd linear ansteigt – ein exzellentes Skalierungsverhalten, das ähnlich gut ist wie das des Loop-Algorithmus. Wenn externe Felder oder chemische Potenziale behandelt werden sollen, ist SSE dem Loop-Algorithmus sogar deutlich überlegen. Außerdem zeigt sich, das SSE wesentlich leichter auf sehr verschiedenartige Hamilton-Operatoren angewandt werden kann als zum Beispiel der Loop-Algorithmus. Die Benchmark-Tests wiesen auch die Leistungsfähigkeit der neuen Messmethode für Greensche Funktionen nach. Am Beispiel der $\langle S^z(\mathbf{r},\tau)S^z(\mathbf{r}',\tau')\rangle$ und $\langle S^z(\mathbf{r},\tau)S^z(\mathbf{r}',\tau')\rangle$ Messungen im Heisenberg-Model wurde ersichtlich, dass die Messung Greenscher Funktionen sogar die bevorzugte Methode sein kann, um auch solche Korrelationsfunktionen zu messen, die diagonal im Ortsraum sind. Anschließend wurden die im Rahmen dieser Arbeit entwickelten Computerprogramme auf verschiedene aktuelle Probleme der Festkörperphysik angewandt. Dies demonstriert sowohl die Leistungsfähigkeit des SSE-Algorithmus als auch die Flexibilität des gewählten Software-Designs, in dem die SSE-Kernalgorithmen, die Modellierung der Gittergeometrie und das physikalische Modell (Teilchen und Wechselwirkungen) in unabhängige objektorientierte Module aufgeteilt wurden. Drei exemplarischen Anwendungen der SSE-Methode schlossen das Kapitel ab: eine Untersuchung zur Zerstörung der suprafluiden und langreichweitigen Ordnung in zweidimensionalen Vielteilchensystemen, eine Arbeit zu Quanten-Phasenübergängen im Hardcore-Bosonenmodell und schließlich eine bosonische Beschreibung der Spin-1/2-Heisenbergleiter, einem Modellsystem für Zink- und Nickelverunreinigungen in Kuprat-Supraleitern.

Der "technische" Teil der Arbeit endete mit einem Kapitel zur Einsetzbarkeit des Computers bei mühsamen algebraischen Termumformungen, wie sie bei der analytischen Behandlung von Vielteilchensystemen entstehen. Diese Überlegungen mündeten in ein Mathematica^{TM}-Programmpaket, das im zweiten Teil des Kapitels 4 für eine analytische Untersuchung des Hubbard-Modells verwendet wurde. Wir studierten dabei die sogenannte Hubbard-I-Näherung und deren Versagen bei Halbfüllung und starker Wechselwirkung,

$U/t \rightarrow \infty$. Es wurde gezeigt, dass die Hubbard-I-Näherung Ladungsfluktuationen auf einem "Hintergrund" von einfach besetzten Plätzen beschreibt. Diese Ladungsfluktuationen sind punktartig und entspechen einem sich zwischen leeren Plätzen bewegenden Elektron sowie einem Elektron, das sich zwischen einfach besetzten Plätzen bewegt. Bei starker Wechselwirkung, also wenn die Hubbard-I-Näherung die Physik des Hubbard-Modells immer schlechter beschreibt, untersuchten wir die wichtigsten Korrekturterme und erzielten mit einer "erweiterten Hubbard-I-Näherung" (EHA) gute Übereinstimmung der analytischen Theorie mit den Ergebnissen numerischer QMC Simulationen. Die punktartigen Ladungsfluktuationen wurden hierbei zu zusammengesetzten Quasiteilchen vergrößert, die zusätzlich zu der Ladungsfluktuation noch eine Spin-, Dichte- oder η-Anregung tragen. In den Einteilchenspektren der EHA-Näherung wird insbesondere die aus den QMC-Daten erhaltene Vier-Band-Struktur korrekt wiedergegeben und erklärt. Weitere QMC-Simulationen, in denen die Spektren der zusammengesetzten Quasiteilchen direkt gemessen wurden, unterstützten die EHA-Interpretation zusätzlich. Insgesamt wurde eine erfolgreiche Methode aufgezeigt, die volle Quasiteilchen-Bandstruktur des Hubbard-Modells zu berechnen – zumindest im paramagnetischen Bereich und bei Halbfüllung.

Der zweite große Teil dieser Arbeit war einer ausführlichen physikalischen Studie gewidmet: der numerischen Analyse des projizierten SO(5)-symmetrischen Modells der Hochtemperatur-Supraleitung. Dieses Modell bildet die niederenergetische Physik der CuO_2-Ebenen in den Kuprat-Supraleitern auf ein System aus vier bosonischen Quasiteilchen-Typen ab: drei Triplett-Anregungen und eine Lochpaar-Anregung. Am Beginn der Untersuchung stand ein kurzer Überblick über die Grundprinzipien der SO(5)-Theorie und der projizierten SO(5)-Theorie der Hochtemperatur-Supraleitung. Danach wurde der Hamiltonoperator eines bosonischen System eingeführt, das die projizierte SO(5)-Symmetrie implementiert: das sogenannte pSO(5)-Modell. Das Modell fällt in die Klasse der zu Anfang dieser Arbeit besprochenen Vier-Bosonen-Modelle, ist also ein effektives Modell für das fermionische t-J-Modell; es ist das einfachste bosonische System, das zwei Schlüsseleigenschaften der Hochtemperatur-Supraleiter enthält: die Mott-Energielücke und das Wechselspiel einer antiferromagnetischen und einer supraleitenden Phase. Numerische Simulationsergebnisse an zwei- und dreidimensionalen Gittergeometrien zeigten, dass das pSO(5)-Modell viele Eigenschaften der Hochtemperatursupraleiter in halbquantitativer oder sogar quantitativer Weise richtig wiedergibt. Insbesondere wurden eine antiferromagnetische und eine supraleitende Phase nachgewiesen, deren Form den entsprechenden Phasen in realen Hochtemperatur-Supraleitern sehr ähnlich ist. Auch die Dotierungsabhängigkeit des chemischen Potenzials und des Resonanzpeaks in Neutronenstreuexperimenten im unterdotierten Bereich werden korrekt beschrieben. Außerdem reproduziert das pSO(5)-Modell Effekte wie Phasenseparation oder die Koexistenz von antiferromagnetischer und supraleitender Ordnung, die in einigen neuartigen Supraleitern auftauchen. Im dreidimensionalen System wurde ein bikritischer Punkt gefunden, an dem die antiferromagnetische und die supraleitende Phase aufeinanderstoßen. Eine Analyse der kritischen Exponenten und Crossover-Parameter an diesem Punkt ergab konsistente Indizien dafür, dass die volle SO(5)-Symmetrie an diesem Punkt in guter Näherung wiederhergestellt wird. Dies ist eine überaus interessante Tatsache, da auf mikroskopischer Ebene im pSO(5)-Modell die volle SO(5)-Symmetrie durch die dem Modell seinen Namen gebende Projektion explizit gebrochen ist.

Zwei ausführliche Anhänge der Arbeit enthalten Dokumentationen und Benutzungshinweise zu den in Rahmen dieser Doktorarbeit entwickelten Alorithmen und Programmpaketen. Der erste Anhang beschreibt die ED-Programme für Hubbard-artige Modelle auf periodische Kristallgittern (`ed_hubb`) und für einzelne Metall-Liganden-Cluster (`ed_clust`), das sehr allgemein einsetzbare SSE-Paket sowie das Computeralgebra-Programm QM. In einem zweiten Anhang wird ein neuer Algorithmus zur effizienten Durchführung selektiver Fourier-Transformationen vorgestellt, der im SSE-Paket verwendet wurde, aber auch unabhängig davon einsetzbar ist.

Lebenslauf

Persönliche Daten	Name	Ansgar Dorneich
	Adresse	Fichtenstraße 7, 71088 Holzgerlingen
	geboren am	4. Juni 1971 in Freiburg
	Familienstand	verheiratet seit 11.09.98, eine Tochter (*26.03.02)
	Staatsangehörigkeit	deutsch
Schulbildung	Sep. 78 – Jul. 82	Grundschule Kirchzarten
	Sep. 82 – Jun. 91	altsprachliches Gymnasium Stegen
	Jun. 91	Abitur
Zivildienst	Jul. 91 – Sep. 92	Mobile Altenbetreuung, Caritasverband Freiburg
	Aug. 92 – Sep. 92	Freiwillige Arbeit in einem Behindertenheim in Yorkshire (England)
Studium	Nov. 92 – Jul. 95	Studium der Physik und Mathematik an der Universität Würzburg
	Aug. 94, Okt. 94	Vordiplom in Mathematik, Vordiplom in Physik
	Aug. 95 – Jul. 96	Graduiertenstudium in Grenoble (Frankreich); Abschluss "Maîtrise de Physique Recherche"
	Apr. 96 – Jul. 96	Forschungspraktikum am Hochmagnetfeldlabor HMFL (Grenoble).
	Okt. 96 – Jun. 97	Beendigung des Studiums in Würzburg
	Jun. 97 – Aug. 98	Diplomarbeit in theoretischer Festkörperphysik bei Prof. Dr. W. Hanke
	Aug. 98	Diplom in Physik
	seit Apr. 99	Aufbaustudium BWL (Fernuniversität Hagen)
Promotion	Okt. 98 – Aug. 01	Doktorarbeit zur Theorie der Hochtemperatursupraleiter bei Prof. Dr. W. Hanke, Institut für Theoretische Physik, Universität Würzburg
Anstellungen	Okt. 98 – Aug. 01	wissenschaftlicher Angestellter an der Universität Würzburg
	seit Sep. 01	Entwicklungsingenieur bei der IBM Deutschland Entwicklung GmbH (Böblingen).
Stipendien, Preise	Mär. 90	1. Preis Bundeswettbewerb Mathematik
	Jun. 91	Preis d. Stiftung „Humanismus Heute"; Preis für das landesbeste Abitur in Baden-Württemberg
	Mär. 93 – Jun. 98	Studienstiftung des deutschen Volkes
	Aug. 98	Diplom in Physik „mit Auszeichnung"
	Dez. 02	W.C. Röntgen-Wissenschaftspreis der Universität Würzburg für die Dissertation

Publikationen

1. W. Hanke, R. Eder, E. Arrigoni, A. Dorneich, S. Meixner, M.G. Zacher:
SO(5) Symmetry in t-J and Hubbard Models,
"Festkörperprobleme/Advances in Solid State Physics" vol. 38,Vieweg Verlag (1999).

2. R. Eder, A. Dorneich, M.G. Zacher, W. Hanke, S.C. Zhang:
Dynamics of an SO(5)-symmetric ladder model,
Phys. Rev. B **59**, 561 (1999).

3. W. Hanke, R. Eder, E. Arrigoni, A. Dorneich, and M.G. Zacher:
SO(5) theory of high-T_c superconductivity: models and experiments,
Proceedings of the Fall 1998 conference Crete, Physica C 317-318, 175 (1999).

4. M.G. Zacher, A. Dorneich, R. Eder, W. Hanke, and S.C. Zhang:
SO(5) symmetry and single particle spectra,
"New Symmetries in Statistical mechanics and Condensed Matter Physics", ed. L. Castellani, Int. J. Mod. Phys. B, Special Issue, vol. 13, No. 23/24 (1999).

5. M.G. Zacher, A. Dorneich, C. Gröber, R. Eder, and W. Hanke:
The Metal-Insulator Transition in the Hubbard Model,
"High Performance Computing in Science and Engineering '99",Springer Verlag (2000).

6. A. Dorneich, M.G. Zacher, C. Gröber, R. Eder:
Strong coupling theory for the Hubbard model,
Phys. Rev. B **61**, 12816 (2000).

7. K. Bernardet, G.G. Batrouni, M. Troyer, A. Dorneich:
Destruction of superfluid and long range order by impurities in two-dimensional systems,
"High Performance Computing in Science and Engineering 2001", Springer Verlag (2002).

8. A. Dorneich, M. Troyer:
Accessing the dynamics of large many-particle systems using stochastic series expansion,
Phys. Rev. E **64**, 66701 (2001).

9. F. Hebert, G.G. Batrouni, R.T. Scalettar, G. Schmid, M. Troyer, A. Dorneich:
Quantum phase transitions in the two-dimensional hardcore boson model,
Phys. Rev. B **65**, 14513 (2002).

10. A. Dorneich, E. Arrigoni, W. Hanke, M. Troyer, and S.C. Zhang:
Phase diagram and dynamics of the projected SO(5) model of high-T_c superconductivity,
Phys. Rev. Lett. **88**, 57003 (2002).

11. A. Dorneich:
A Fast algorithm for selective Fourier transformations,
submitted to Comp. Phys. Comm.

12. A. Dorneich, W. Hanke, E. Arrigoni, M. Troyer, and S.C. Zhang:
Dynamical properties and the phase diagram of the projected SO(5)-symmetric model of high-T_c superconductors,
to appear in the Journal of Physics and Chemistry of Solids for the proceedings of the ISSP8 symposium on Correlated Electrons (2001).

13. G. Schmid, S. Todo, M. Troyer, A. Dorneich:
Finite-temperature phase diagram of hard-core bosons in two dimensions,
Phys. Rev. Lett. **88**, 167208 (2002).

14. K. Bernardet, G.G. Batrouni, J.-L. Meunier, G. Schmid, M. Troyer, and A. Dorneich:
Analytical and numerical study of hardcore bosons in two dimensions,
Phys. Rev. B **65**, 104519 (2002).

15. K. Bernardet, G.G. Batrouni, M. Troyer, A. Dorneich:
Destruction of superfluid and long range order by impurities in two-dimensional systems,
"High Performance Computing in Science and Engineering 2001", Springer Verlag (2002).

16. A. Dorneich, M. Jöstingmeier, E. Arrigoni, C. Dahnken, T. Eckl, W. Hanke, S.C. Zhang, M. Troyer:
An object oriented C++ class library for many-body physics on finite lattices and a first application to high-temperature superconductivity
To appear in the proceedings of the First Joint HLRB and KONWIHR Result and Reviewing Workshop, Garching, Oct. 2002.

17. A. Dorneich, E. Arrigoni, M. Jöstingmeier, W. Hanke, S.C. Zhang:
Scaling properties of the projected SO(5) model in three dimensions
will be submitted to Phys. Rev. B.

Danksagung

An erster Stelle möchte ich mich bei Herrn Professor Dr. Hanke bedanken, der mich in den letzten Jahren an mehrere hochinteressante und aktuelle Themen aus dem Bereich der Vielteilchenphysik herangeführt hat und es mir gleichzeitig ermöglichte, meine Arbeitsschwerpunkte sehr selbständig zu wählen.
Mein besonderer Dank gilt auch Herrn Professor Dr. Matthias Troyer (ETH Zürich), mit dem mich in den letzten zwei Jahren eine menschlich sehr angenehme und fachlich für mich außerordentlich fruchtbare Zusammenarbeit verbunden hat. Von seinem einzigartigen Wissensschatz auf fast allen Teilgebieten des Scientific Computing hat meine Arbeit enorm profitiert.
Herrn PD Dr. Robert Eder war seit meiner Diplomandenzeit ein ständiger Begleiter und Betreuer meiner Arbeit. Seine kameradschaftliche Art, sein immer offenes Ohr, sein Understatement und seine Fähigkeit, aus mathematischen Formalismen klare physikalische Bilder herauszuarbeiten und neue Ideen zu entwicklen, haben die Zusammenarbeit zu einem Vergnügen gemacht. Die thematisch vielfältigen gemeinsamen Publikationen zeugen davon.
Von Herrn PD Dr. Enrico Arrigoni konnte ich während der Arbeit am pSO(5)-Modell sehr viel lernen. Für die große Geduld und Freundlichkeit, mit der er mich an seinem umfangreichen Wissen in Theoretischer Physik teilhaben ließ, möchte ich herzlich danken.
Bei Herrn Prof. Dr. S.C. Zhang (Stanford University) bedanke ich mich für seine vielen Anregungen und die angenehme Zusammenarbeit am pSO(5)-Modell.
Herrn Dr. Marc Zacher danke ich für die Einarbeitung in das Maximum-Entropy-Programm, für die langjährige gute Zusammenarbeit bei Rechenzeitanträgen, Berichten und Publikationen und nicht zuletzt für die gemeinsamen Stunden im Tiefschnee.
Herzlichen Dank an meinen Zimmernachbarn Herrn Thomas Eckl für die gute gemeinsame Zeit, viele Diskussionen und dafür, dass er meine Marotten wie zum Beispiel den geräuschvollen Verzehr von Crackern[1] jeden Tag zwischen 16 und 18 Uhr klaglos ertragen hat.
Herrn Martin Jöstingmeier danke ich für die gute Zusammenarbeit während dessen Diplomarbeit und für seine Souveränität in zahlreichen weltanschaulich-politisch-philosophischen Debatten, in deren Verlauf er selbst ironisch-unsachliche Kommentare gelassen konterte oder wegsteckte.
Allen Würzburger Kollegen, insbesondere Berthold Brendel, Chris Dahnken, Dr. Robert Eder, Thomas Eckl, Carsten Gröber, Martin Jöstingmeier, Tobias Mensing, Christine Schmeisser und Dr. Marc Zacher, danke ich für die tolle Atmosphäre in der Arbeitsgruppe. Diese Atmosphäre erreichte meist bei den gemeinsamen Mensabesuchen und Kaffeepausen ihren Höhepunkt und brachte Gespräche von unerhörter Tiefgründigkeit und kristallener Klarheit des Gedankens zu beinahe jedem denkbaren Thema hervor.
Der Familie Kemmer in Würzburg danke ich für die freundliche Aufnahme in ihr Haus, das mir in den letzten Jahren eine ideale 'Forscherklause' bot, sowie für das kameradschaftliche Verhältnis zu mir und den anderen Untermietern.
Meine Frau Astrid hat mich in den letzten drei Jahren von Montag bis Freitag vollzeit

[1]Ritz-Cracker, 200g Packung

und am Wochenende in Teilzeit für die Physik freigestellt. Sie hat mir an den gemeinsamen Wochenenden in Asperg Entspannung, Erholung, gutes gemeinsames Essen und viel gemeinsames Lachen ermöglicht und (mehr oder weniger) gutmütig meine Neigung toleriert, innerhalb kürzester Zeit nach meiner Ankunft große Teile der kleinen gemeinsamen Wohnung mit diversen wichtigen Papieren zu übersähen. Damit hat sie immens zum Gelingen dieser Arbeit beigetragen.
Schließlich danke ich meinen Eltern für alles, was sie für mich getan haben.